COMPUTER CONTROL
OF INDUSTRIAL
PROCESSES

Editors S.Bennett & D.A.Linkens

COMPUTER CONTROL OF INDUSTRIAL PROCESSES

Editors S.Bennett & D.A.Linkens

PETER PEREGRINUS LTD.
On behalf of the
Institution of Electrical Engineers

Published by: The Institution of Electrical Engineers, London
and New York
Peter Peregrinus Ltd., Stevenage, UK, and New York

© 1982: Peter Peregrinus Ltd.

British Library Cataloguing in Publication Data

Bennett, S.
  Computer control of industrial processes.—
  (Control engineering; 21)
  1. Process control—Data processing
  2. Automatic control—Data processing
  I. Title   II. Linkens, D.A.
  III. Series
  670.42'7   TS156.8

ISBN 0 906048 80X

Printed in England by Short Run Press Ltd., Exeter

Contents

Preface   ix

Editor's Introduction   xi

1. **A Survey of Computer Control**
   1.1 Introduction 1
   1.2 The elements of a computer control system 2
        1.2.1 Measurement 2
        1.2.2 Actuation 4
        1.2.3 Direct digital control (DDC) 4
        1.2.4 Sequence control 7
        1.2.5 Supervisory control systems 8
        1.2.6 The engineer and the process control computer 9
        1.2.7 Facilities for plant operator and plant manager 10
   1.3 Hardware for computer control systems 11
   1.4 Integrity of computer control systems 14
   1.5 The economics of computer control systems 15
   1.6 Future trends 16
   1.7 Conclusions 16
   1.8 Bibliographic notes 17

2. **Process Control by Computer**
   2.1 Introduction 18
        2.1.2 Evolution of computer control 18
        2.1.3 Supervisory control 19
        2.1.4 Direct digital control 19
   2.2 Typical control schemes: hardware features 21
        2.2.1 Data input methods 21
        2.2.2 Data output methods 22
        2.2.3 Interface control 22
        2.2.4 Emergency back-up 24
   2.3 Software 26
   2.4 Integral control and bumpless transfer 28
        2.4.1 At the back-up station 28
        2.4.2 Software for bumpless transfer 29
        2.4.3 General 30
   2.5 Integrity of interactive systems 31
        2.5.1 Theoretical background 31
        2.5.2 Distillation column example 33
   2.6 Conclusion 37
   2.7 References 37

3. **Some DDC System Design Procedures**
   3.1 Introduction 38
   3.2 Choice of sampling frequency 39
   3.3 Frequency domain compensation method 41

3.4   The compensation of class 0 (regulator) systems 47
3.5   Noisy input or output signals 48
3.6   Structural resonances and digital notch networks 48
3.7   Coefficient quantisation in a discrete controller 49
3.8   Arithmetic roundoff-noise in a discrete controller 51
3.9   Multirate and subrate controllers 52
3.10 Time domain synthesis with polynomial inputs 53
3.11 Comparison of time domain synthesis and frequency domain compensation techniques 57
3.12 Conclusions 58
3.13 References 59
      APPENDIX 3.1 List of symbols 59

4. Self-tuning Digital Control Systems
    4.1   Introduction 61
    4.2   Self-tuning sequence 62
    4.3   System identification 64
    4.4   Controller synthesis 66
    4.5   Basic self-tuning property [1, 2, 3, 4] 75
    4.6   Time-varying parameters and time-delays 77
    4.7   Multivariable self-tuners 77
    4.8   Extended self-tuning algorithm [52] 78
    4.9   Non-parametric self-tuners 79
    4.10 Self-tuning prediction [29] (see also Appendix 4.1) 80
    4.11 Applications 80
    4.12 Further topics 80
    4.13 Concluding remark 81
    4.14 References 82
          APPENDIX 4.1 Self-tuning prediction and recursive approximate maximum likelihood 85
          APPENDIX 4.2 List of symbols 86

5. Requirements for Real-Time Computing
    5.1   Introduction 88
          5.1.1   Central processing unit 89
          5.1.2   Storage 90
          5.1.3   Input and output 90
    5.2   Interrupts 92
    5.3   Characteristics of interrupt systems 95
          5.3.1   Interrupt inputs 95
          5.3.2   Interrupt response 95
          5.3.3   Saving and restoring registers 96
          5.3.4   Determining interrupt sources 97
          5.3.5   Priority 99
          5.3.6   Enabling and disabling interrupts 102
    5.4   Operating systems 102
          5.4.1   Operating system components 103
          5.4.2   Overview 105
          5.4.3   Supervisor calls 105
          5.4.4   Memory management 106
          5.4.5   Spooling 108
          5.4.6   Multi-terminal moniter 110
    5.5   Conclusion 110
    5.6   References 110

6. Communications for Distributed Control
    6.1   Introduction 112
    6.2   Advantages of distributed control 112
    6.3   Network configurations 114
          6.3.1   Network classification 114
          6.3.2   Example configurations 114
          6.3.3   A comparison of store and forward and broadcast configurations 118
          6.3.4   Interconnected networks 120
          6.3.5   Station structure 120

6.4   Layered network structure 121
      6.4.1   Components of a layered structure 121
      6.4.2   Reasons for layered structure 123
      6.4.3   A layered network architecture for process control 124
6.5   Network management 127
      6.5.1   Station management function 127
      6.5.2   Network control centre functions 128
6.6   Standards activity 128
6.7   References 130

7.   **Languages for Computer Control**
7.1   Introduction 131
7.2   CORAL 66 and RTL/2 132
7.3   MODULA 132
7.4   ADA 135
7.5   References 136

8.   **Online Computer Control of pH in an Industrial Process**
8.1   Introduction 137
8.2   Process 139
8.3   Computer installation 144
8.4   Investigations 145
      8.4.1   Introduction 145
      8.4.2   Jacketed PID control (stage b) 147
      8.4.3   Jacketed adaptive control (stage c) 148
8.5   Conclusions 151
8.6   Acknowledgments 152
8.7   References 152
      APPENDIX 8.1 List of symbols 154

9.   **Direct Digital Control in CEGB Power Stations**
9.1   Introduction 155
9.2   The cost/benefits of DDC 156
      9.2.1   Reliable information displays 156
      9.2.2   Less maintenance 156
      9.2.3   Improved control 157
9.3   Scope of the boiler controls 158
      9.3.1   Load and pressure control 158
      9.3.2   Feedwater control 159
      9.3.3   Steam temperature control 159
      9.3.4   Combustion air controls 159
9.4   Structures for load and pressure control 159
      9.4.1   Plant interactions 160
      9.4.2   Boiler – follows – turbine control 161
      9.4.3   Turbine – follows – boiler control 161
      9.4.4   High security control 162
9.5   Engineering a DDC system 163
      9.5.1   Distribution of the control functions 163
      9.5.2   Computers 164
      9.5.3   Interface equipment 164
      9.5.4   Control desk 165
      9.5.5   Plant modifications 165
9.6   The CUTLASS software 165
      9.6.1   CUTLASS requirements 166
      9.6.2   General features of CUTLASS 166
      9.6.3   Features of the modulating control language subset 167
9.7   Conclusions 168
9.8   Acknowledgments 168
9.9   References 168

10 Structured Analysis of Manufacturing Systems
10.1 Introduction 170
10.2 Manufacturing industry 171
    10.2.1 General 171
    10.2.2 Manufacturing systems 171
    10.2.3 Activities and communications 172
    10.2.4 Types of production 173
    10.2.5 Mechanisation and automation 174
    10.2.6. Numerical control 175
    10.2.7 Computer-aided manufacture 176
    10.2.8 Integrated computer-aided manufacturing systems 178
10.3 Functional and information models of a typical Batch manufacturing system 179
    10.3.1 Structured analysis 179
    10.3.2 Illustrative example 181
10.4 Acknowledgments 183
10.5 References 183

11. In-Process Surface Measurement and the Computer Control of Manufacturing Processes
11.1 Why control? past and present 196
11.2 Control what? 197
11.3 Some modern methods of control 200
11.4 Conclusions 204
11.5 References 205

Preface

PREFACE

    This text contains the main lectures given at a Vacation School on Computer Control of Industrial Processes held at the University of Sheffield in September 1980. The purpose of the School, one of a series sponsored by the Science and Engineering Research Council, was to provide a survey of the current work in Computer Control to research students working in British Universities on control topics and young graduates in industry. In addition to the major lectures, which are reproduced here, there were revision lectures covering computer hardware and sampled data theory, laboratory classes and case studies dealing with specific control problems encountered in a range of industries.

    As organisers of the school, as well as editors of this book, we wish to take the oppotunity of thanking all contributors and for their contributions to the School and this book. Particular mention must also be made to the clerical and technical staff of the Department of Control Engineering, University of Sheffield, who before and during the School provided the background support vital for its success. Also we are indebted to Miss Deborah Macdonald for producing the typescript from which this book has been printed.

S. Bennett
D.A. Linkens

IEE CONTROL ENGINEERING SERIES 21

SERIES EDITORS: Prof. B H Swanick
Prof. H Nicholson

CHAPTER AUTHORS

1. **A Survey of Computer Control**
   Dr. D J Sandoz, Electrical Engineering Laboratories, University of Manchester

2. **Process Control by Computer**
   J B Edwards, Department of Control Engineering, University of Sheffield

3. **Some DDC System Design Procedures**
   Dr. J B Knowles, UKAEA, Control & Instrument Division, Atomic Energy Establishment, Winfrith

4. **Self-tuning Digital Control Systems**
   Dr. P E Wellstead, Control Systems Centre, UMIST, Manchester

5. **Requirements for Real-Time Computing**
   Prof. M J H Sterling, Department of Engineering Science, University of Durham

6. **Communications for Distributed Control**
   Dr. M S Sloman, Department of Computing & Control, Imperial College of Science & Technology, University of London

7. **Languages for Computer Control**
   Prof. I C Pyle, Department of Computer Science, University of York

8. **Online Computer Control of pH in an Industrial Process**
   Dr. O L R Jacobs, Department of Engineering Science, Oxford University
   Dr. P Hewkins, Brown Boveri & Cie, Heidelberg, Dr. C While, ICI Ltd, Huddersfield

9. **Direct Digital Control in CEGB Power Stations**
   Dr. M L Bransby, CEGB, North Eastern Region, Scientific Services Department Harrogate

10. **Structured Analysis of Manufacturing Systems**
    Prof. T R Crossley, Department of Mechanical Engineering, University of Salford

11. **In-Process Surface Measurement and the Computer Control of Manufacturing Processes**
    Prof. D J Whitehouse, Department of Engineering, University of Warwick

Editors introduction

The use of automatic feedback control in the process industries began in the 1930's with the introduction of pneumatic and hydraulic two and three-term controllers. By 1960 a wide range of electronic, pneumatic and hydraulic controllers had become available, since then there has been a changeover to the use of computers for process control and in particular the impact of the microprocessor over the last five years has been such that the use of the microprocessor has become the norm. In Chapter 1, Sandoz outlines the developments since 1960 and briefly describes the control requirements in the process industries.

The design of feedback controller algorithms can be approached either from noise-free deterministic considerations or noise-contaminated stochastic methods. Also, design may concentrate on a single loop structure or it may consider an interacting multi-loop environment. Each of these approaches can be found in the Chapters of this book which deal with the design of feedback regulator algorithms.

The most commonly employed feedback algorithm found in process control is that of the PID regulator. This is very widely used in single and multi-loop applications particularly where little is known about the process dynamics. The tuning of the PID coefficient is a common 'art' amongst process control engineers. In Chapter 1 direct digital control (DDC) using a difference equation equivalent to the classic analogue three-term controller is described. Although integral control is normally essential to remove offset from the set-point in the plant output, problems due to 'integral wind up' caused by actuator saturation are well known and are mitigated using 'integral desaturation' schemes. These concepts are further described and illustrated with a distillation column example by Edwards in Chapter 2 which also considers the requirement for 'bumpless transfer' when switching from manual to automatic feedback control. The majority of computer control schemes use DDC concepts of this nature with interaction between loops via cascade connections of set-points.

Apart from on-line tuning of PID controller parameters a wealth of knowledge exists for the classical design of single-loop systems based on deterministic conditions. The extension of methods due to Nyquist, Bode etc., for continuous analogue systems to discrete digital systems presupposes some knowledge of z-transforms. In the Summer School on which this book is founded, a number of introductory lectures were given including one on sampled-data theory. This material has been excluded from the book since it is readily available in many standard texts on control systems. Knowles, in Chapter 3, discusses a number of simple techniques which facilitate the conversion of continuous control concepts to digital computer application. Choice of sampling frequency is fundamental, while limitations due to coefficient quantisation and arithmetic roundoff-noise are clearly important. The design of digital compensators is mainly approached in the frequency domain using classical servomechanism techniques suitably adapted for discrete implementation. Brief consideration is also given to filter rejection of unwanted signals due to noise or resonances, and to the use of multi-rate sampling to reduce actuator wear and output ripple.

If Chapters 1 to 3 represent the intuitive and classical approaches to controller design, then Wellstead in Chapter 4 represents the current interest in self-adaptive control for industrial processes. The widespread use of tuned PID regulators underlines the twin requirements of the process control industry - those of the ability to cope with unknown process dynamics and the need for 'robust' controller algorithms. The current interest in self-tuning control majors on these two concepts. The two schools of thought in self-tuning control are commonly referred to as 'implicit' and 'explicit' protagonists. In the implicit self-tuner the controller parameters are estimated directly without estimation of the process dynamics in contrast to the explicit approach which first estimates the plant dynamics and subsequently solves an identity for determination of the controller parameters. The origins of the implicit approach are found in optimal control theory and stochastic regulation, whereas the explicit method stems from classical deterministic servomechanism design methods. In Chapter 4 a wide-ranging review of many aspects of self-tuning is given, including the basic principles and underlying identification strategies. Particular attention is given to the requirement of tracking changing input demands (the servo problem) as well as maintaining constant ouput levels (the regulator problem). Extensions of the self-tuning principle to the multivariable case and prediction problems are covered and a brief survey given of applications where self-tuning has been applied.

One application of self-tuning is covered in detail by Jacobs in Chapter 8, where a number of pilot studies in on-line computer control of pH in an industrial process are described. The plant studied is a continuous stirred-tank reactor and is of particular interest because of the highly non-linear nature of the pH characteristic. In this application comparisons were made between classical PID control and a self-tuner of the implicit variety. In both cases linearisation was obtained using an inverse pH characteristic. The results show the advantages that can be obtained using modern adaptive computer control algorithms, and also illustrate the widely-recognised fact that self-tuning can sometimes 'blow-up' and must therefore have alarm limits built in to it so that switch over to lower performance algorithms such as PID can be made at crucial times.

The theory and application of multivariable self-tuning regulators is new and rapidly emerging, and will doubtless require a considerable development period. A particular application of multivariable control approached from a different angle is given by Bransby in Chapter 9, where computer control of a coal-fired drum boiler in a power station is considered. A number of control areas are allocated each comprising several loops and each requiring its own integrity. The load and pressure control area is highlighted and plant interactions indicated via a linearised transfer function matrix. Based on knowledge of plant behaviour and requirements, three possible controller strategies are considered encompassing classical PID concepts and modern multivariable analysis techniques.

The reduction in cost of computer hardware has made it feasible to consider multiple computer systems as a means of simplifying design and increasing reliability. The most commonly used schemes have been stand-by systems with either manual or automatic changeover to a second computer. The advent of the microprocessor has brought to the fore consideration of distributed computer control systems. Such schemes involve communication links and in Chapter 6 Sloman discusses in detail techniques for inter-computer communications.

The weakest area in ensuring reliable operation of computer control schemes is that of software design and implementation. It has only recently been recognised that real-time programming is significantly more difficult than ordinary sequential programming. Computer control systems involve concurrency (i.e. several programmes apparently running at the same time), they have to run continuously, actions have to be performed within specified times and access to special input/output facilities is required. The development of high level languages greatly eased the problem of program design and implementation for ordinary sequential programming, but these programmes do not provide access to interrupts, do not provide the ability to define sections of programme as processes to be run concurrently, do not provide facilities to programme exception activities (i.e. actions to be performed on detection of a run-time error).

In the early process control applications access to such facilities was obtained by writing the software entirely in assembler language and some users still adopt this approach. Users of this approach, in order to provide reliabilty, restrict themselves in subsequent designs to re-using existing modules and limiting new coding to a minimum. The major disadvantage is that in a time of rapid hardware development, a user may become heavily dependent on obsolete and expensive hardware. By the mid-1960's some of the companies involved in process control realized that similar facilities were required in many applications and began to produce operating systems specifically for real-time operation. Sterling in Chapter 5 discusses the requirements of a real-time computer system of this type.

The user now had a choice, he could continue to programme in assembler language, making calls to various operating facilities or he could programme in a high level language (again making calls to use operating system facilities). Various Fortran-based languages - usually called Process Fortran - began to appear, the best of which provided the programmer with a wide range of facilities, often enabling assembler coding to be freely mixed with the Fortran code. An alternative to this approach was the development of interpreters which allowed the process engineer rather than the specialist programmer to implement a control scheme. In either case the application software is not portable and with interpreters or large general purpose operating systems considerable time overheads can be incurred. For many applications there is a need for a language in which real-time operations can be directly expressed without the requirement of an extensive operating system, in this way only the facilities actually required for the application will be present in the software. In Chapter 7, Pyle describes some of the currently available languages and briefly discusses the reasons behind the huge investment being made in ADA.

An important feature of ADA, as well as its strong typing and provisions for concurrency, is the provision for handling EXCEPTIONS. Real-time systems have to run continuously and the designer has to attempt to forsee all possible events which could cause the software to crash. In many real-time systems a considerable amount of the programming effort is devoted to providing checks on data to prevent the generation of run-time errors (with the consequence of slower operation). One of the features of CUTLASS, described by Bransby in Chapter 9 is the flagging of bad data and one of the attractions of many BASIC interpreters is the ability to provide for exceptions by the use of the ON ERROR GOTO statement. In ADA the exception feature is taken further in that exceptions can be raised either automatically through run-time errors detected by the operating system or directly by the programmer. The latter facility enables the programme designer to clearly distinguish between normal and abnormal operations, and is but one example of the move towards software engineering.

Closely associated with software engineering have been the development of design languages/techniques for complex organisational systems. Crossley, in Chapter 10, describes the application of one such technique to the analysis of a batch manufacturing company. These design languages are functionally similar to the design techniques described in Chapters 3 and 4, they provide the engineering designer with the tools which he can use to formulate in precise terms a control algorithm or a system model.

Despite the fact that the design techniques are expressed in precise mathematical language the engineering designer has to exercise perception in the choice and use of a design technique. In a similar manner the software designer although constrained in a precise way by the structure of the language used has considerable freedom, in fact many would agree that he has too much freedom. So much so that many of the developments in software engineering have been towards limiting that freedom.

Chapter 1

A survey of computer control

1.1 INTRODUCTION

The first use of computers to control industrial processes occurred in the early 1960's. There is, inevitably, a conflict of claims as to which country and company made the first application. In Britain, ICI led the way with an application on a chemical plant at Fleetwood in Lancashire. A Ferranti Argus 200 computer was used. This computer was programmed by physically inserting pegs into a plug board, each peg representing a bit in a computer memory word (it proved more reliable than its early successors because the destruction of memory contents required the dislodgement of pegs rather than the disruption of the magnetic status of ferrite cores). A British computer manufacturer in league with a British chemical company was amongst the first pioneers of the application of computers for process control.

The situation has changed. The technology of both hardware and software has progressed rapidly. The British computer industry has lost its early initiative and the greater part of progress today is motivated by developments from the United States. The biggest impact of all has come from the micropocessor. Computer control systems, once prohibitively expensive, can now be tailored to fit most industrial applications on a competitive economic basis. The use of microprocessors for process control is becoming the norm.

These advances have motivated many changes in the concepts of the operations of industrial processes. Video display terminals now provide the focus for operators to supervise plant. Large panels of instruments, knobs and switches are replaced by a few keyboards and screens. Control rooms are now much smaller and fewer people are required to supervise a plant.

Process computers now have the capability to implement sophisticated mathematical analysis to aid effective operation. Plant managers and engineers can be provided with comprehensive information concerning the status of plant operations. This motivates more effective overall management of process plant. Surprisingly, and in contrast, the concepts of basic feedback control implemented by the computers have changed little from the days when pneumatic instrumentation was the main means for implementation. Direct digital control is essentially a computer implementation of techniques that have long been established as standard for industrial process control.

This chapter reviews the main functions of a process control computer, including measurement and actuation, direct digital control, sequence control, supervisory control and operator communications. The hardware and software aspects of distributed and hierarchical computer control systems and the integrity and economics of computer operations are discussed. The intention is to provide a broad introductory overview of the significant features of computer control systems.

1.2 THE ELEMENTS OF A COMPUTER CONTROL SYSTEM

This section describes the main tasks that computer control systems perform. Fig. 1.1 is a broad illustration of a computer control system. Signals are monitored from and are supplied to the controlled process. Collected data are analysed at various levels to provide plant adjustments for automatic control or to provide information for managers, engineers and operators.

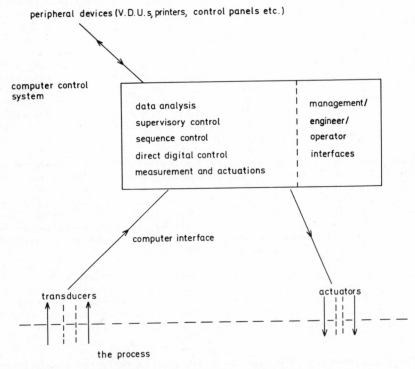

Fig. 1.1 Schematic diagram of a computer control system

1.2.1 Measurement

The plant operation is monitored using transducers. These are devices that generate an electrical signal that is proportional to a physical quantity on the plant that is to be measured (i.e. temperature, pressure, flow, concentration etc.). The transducer signals are connected to the measurement interface of a computer system, which is usually standardised to accept signals of one particular kind. A widely adopted standard is that the 0 to 100% range of any particular measurement converts to 4 to 20mA or 1 to 5V. For example if a particular transducer monitors flow over the range 10 litres/hr to 50 litres/hr, then the transducer would be calibrated so that 10 litres/hr supplies 4mA and 50 litres/hr supplies 20mA to the computer interface. A different standard may sometimes be adopted for temperature measurements taken via thermocouple or resistance thermometer transducers.

Another variety of measurement concerns the status of various aspects of the controlled process. Is a valve open or closed? Is a vessel full of liquid? Is a pump switched on? Such information would be supplied to the computer in digital form, perhaps by the opening and closing of relay switches or by the level of a TTL voltage.

The computer may also monitor digital data directly, via a serial or parallel data communications link. Many transducers now utilise microprocessors, for example liquid concentration analysers. Typically a microprocessor might apply statistical analysis to extract the required information from the monitored plant signal. Given a numerical result it is then straightforward to effect the transfer to the control computer.

The control computer maintains a record of all of the measurements. Periodically this record is updated by scanning each of the signals connected to the interface. Each measurement may be referenced, say, twice each second. At these instances, the electrical signal is converted to a numerical equivalent by an analogue-to-digital converter (ADC). If the ADC converts to a 10-bit binary number, which is commonly the case, then 4mA returns 0 and 20mA returns 1023 for the standard signal. To be meaningful, the returned number must be scaled to engineering units. For a transducer with a range of 10 to 50 litres/hr, 0 scales to 10 and 1023 to 50. An intermediate value y ($0 < y < 1023$) scales to $(y/1023) \times 40 + 10$. The incoming signal sometimes has to be linearised; certain flow transducers require the square root of a signal to be evaluated after measurement; thermocouples generate a voltage that relates to temperature via a polynomial formula which has to be evaluated after measurement.

If a measurement signal has a lot of noise superimposed upon it, digital filtering may be applied to smooth the signal. In such a case, if a representative value is required every second, then the signal must be sampled and processed by the filter more frequently (say every 0.1 secs). The most commonly used filtering procedure employs the first order exponential algorithm

$$yf_{k+1} = a.yf_k + (1-a)y_{k+1} \qquad k = 0, 1, 2 \ldots$$

where yf_k is the filtered value at instant k, y_{k+1} is the measurement at instant k+1 and $a = \exp(-T/\tau)$ with T = the interval k to k+1 (0.1 secs) and τ is the filter time constant. The selection of the times T and τ depends upon the frequency at which the filtered measurement is wanted (1 second in this example) and the frequency and amplitude of the noise on the measurement.

Most measurements taken by the process control computer will also be checked to ascertain whether or not the plant is in a safe state of operation. Two sets of limits usually relate to each measurement. If the inner range is exceeded, a warning status is established. If the outer range is exceeded then a 'red alert' condition applies. Automatic checking for such emergency conditions is a very important feature of the on-line computer.

Therefore, for on-line control, each measurement must be converted, possibly filtered, possibly linearised, scaled to engineering units and, finally, checked against alarm limits. The intervals between measurement samples may be quite long in some cases (many seconds) but in other cases many samples may have to be taken each second. If there are a lot of different measurements to be taken, and some installations require many hundreds, there is considerable data processing necessary to bring all of the measurements to a form meaningful for inspection by humans and for use in controllers. Careful choice of sampling intervals, e.g. by not monitoring signals more frequently than necessary, can reduce the computational burden. The selection of such intervals also requires care to ensure that problems such as signal aliasing do not arise.

1.2.2 Actuation

Control of plant is usually achieved by adjusting actuators such as valves, pumps and motors etc. The control computer may generate a series of pulses to drive the actuation device to its desired setting. In such a case, the drive signal would be generated as a relay contact closure or a change in voltage level. Alternatively, a voltage that is proportional to a desired setting may be produced by a digital-to-analogue converter (DAC). An actuation device will often supply a measurement back to the computer so that it is possible to check whether or not an actuation command has been implemented. The computation associated with actuation is usually small, however, some pulse driven actuators require frequent associated measurement to determine when the desired setting has been reached. In this case a more significant computational burden can be incurred.

1.2.3 Direct Digital Control (DDC)

Conventional analogue electronic control systems employ the standard three term algorithm

$$u = K_p(e + \int \frac{edt}{T_i} + T_d \cdot \frac{de}{dt}) \tag{1.1}$$

with e = r −y, y is the measurement, r is the reference or set point, e is the error, K_p is the controller gain ($1/K_p$ is the proportional band), T_i is the integral action time and T_d the derivative action time. There are variations on this form. Very few controllers actually use derivative action. If it is used de/dt is sometimes replaced by dy/dt to avoid differentation of set point. Nearly all industrial control problems are solved by application of this algorithm or close variations to it.

Most DDC systems utilise a difference equation equivalent to the above algorithm. If the computer recalculates the actuation signal u every T seconds, then the most simple numerical approximation employs

$$\frac{de}{dt} \simeq \frac{e_k - e_{k-1}}{T} \quad \text{and} \quad \int e.dt = \sum e_k.T \qquad k=0,1,2,\ldots \qquad (1.2)$$

with the interval k to k+1 equal to T seconds.

The control equation then results as

$$u_k = K_p \{e_k + \frac{T.s_k}{T_i} + \frac{T_d}{T} (e_k - e_{k-1})\} \qquad (1.3)$$

with $s_k = s_{k-1}+e_k$ being the sum of errors.

The advent of cheap microprocessors has made it very simple to programme and implement this control equation. However, there are many traps into which the self-taught control engineer can fall. One of the most significant is associated with saturation of the actuation signal. This signal must be considered to lie within a defined range Umin/Umax. If the control signal saturates at either extreme careful consideration has to be given to the integral sum s_k at this stage. If the summation procedure were to continue unchecked s could take up a large and unrepresentive value which could lead to much degraded control system performance thereafter. Special procedures for integral desaturation have been developed to accompany the controller. These ensure that the actuation signal emerges from saturation at a timely moment so that an effective control system response is generated.

DDC may be implemented on a single loop basis by a single microprocessor controller or by a larger computer which could implement upwards of a hundred control loops. The total control system for an industrial process can become quite complex. DDC loops might be cascade connected, with the actuation signals of particular loops acting as set points for other loops. Signals might be added together (ratio loops) and conditional switches might alter signal connections. Fig 1.2 is a diagram of a control system that has recently been implemented by a computer on an ICI plant. The pressure of steam produced by a boiler is to be controlled. This is achieved by regulating the supply of fuel oil to the burner. However, a particular mix of fuel and air is required to ensure efficient and non-polluting combustion. The purpose of this illustration is not to discuss how the control system achieves its objective, but rather to indicate some of the elements that are required for industrial process contol. Such elements must be available as features of a process control computer.

Referring to Fig. 1.2 the steam pressure control system generates an actuation signal that is fed to an automatic/manual bias station. If the latter is selected to be in automatic mode, the actuation signal is transmitted through the device; if it is in the manual bias mode, the signal that is transmitted is one that has been manually defined (e.g. by typing in a value on a computer keyboard). The signal from the bias station is connected to two units, a high signal selector and a low signal selector. Each selector has two input signals and one output. The high selector transmits the higher of its two input signals and the low selector the lower of its two input signals. The signal from the low selector cascades a set point to the DDC loop that controls the flow of oil. The signal from the high selector cascades a set point to the DDC

loop that controls the flow of air. Finally, a ratio unit is installed
in the air flow measurement line. A signal that is generated from
another controller is added to the air flow signal prior to its being
fed to the air flow controller. This other controller is one that
monitors the combustion flames directly, using an optical pyrometer for
measurement, and thereby obtains a direct measure of combustion
efficiency.

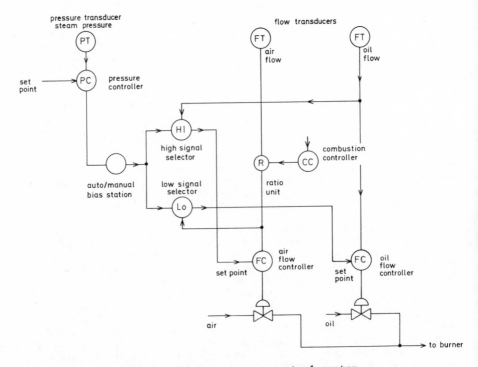

Fig. 1.2 Boiler pressure control system

Another controller that is of value but is not a feature of Fig. 1.2
is one based on the lead/lag transfer function. This can be used to
provide feed-forward compensation so that the effect of disturbances
upon the plant, for example because of an alteration in the
environmental conditions, is minimised.

Hence there is much more to computer control than simple DDC. The
boiler example is probably more complex than many industrial schemes
although the use of additional signal processing over and above simple
DDC is very common.

DDC is not restricted to the three-term algorithm described above,
although the latter is almost universally used in the process
industries. Algorithms based upon, for example, z-transform design
techniques can be equally effective and a lot more flexible than the
three-term controller. However, the art of tuning three-term controllers
is so well established amongst the control engineering fraternity that

new techniques are slow to gain ground. The fact that three—term control copes perfectly adequately with 90% of all control problems is also a deterrent to general acceptance of new concepts for control system design.

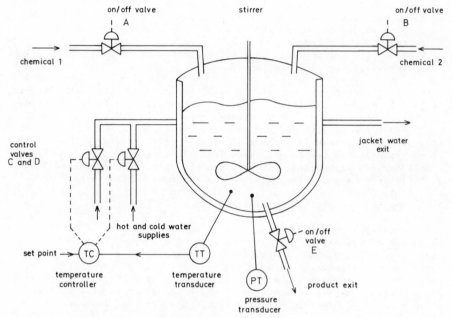

Fig. 1.3 A chemical reactor

1.2.4 Sequence Control

Many industrial processes are required to be automatically sequenced through a number of stages during their manufacturing operations. For example, consider the manufacture of a chemical that is produced by reacting together two other chemicals at a particular temperature. Fig.1.3 illustrates a typical plant arrangement for this purpose. The chemicals react in a sealed vessel (the reactor). The contents are temperature controlled by feeding hot or cold water to the water jacket that surrounds the vessel. This water flow is manipulated by adjusting the control valves C and D. On/off valves A, B and E in the chemical supply and vessel exit pipelines are used to regulate the flow of material into and out of the vessel. The temperature of the vessel contents and the pressure at the bottom of the vessel are monitored.

The manufacturing procedure for this plant might involve the following stages of operation.

1) Open valve A to charge chemical 1 to the reacting vessel.
2) Check the level of chemical in the vessel (by monitoring the pressure vessel). When the required amount of chemical has been charged, close valve A.
3) Start the stirrer to mix the chemicals together.
4) Repeat stages 1 and 2, with valve B, in order to charge the second

chemical to the reactor.
5) Switch on the three-term controller, and supply a set point so that the chemical mix is heated up to the required reaction temperature.
6) Monitor the reaction temperature. When it has reached set point, start a timer to time the duration of a reaction.
7) When the timer indicates that the reaction is complete, switch off the controller and open valve C to cool down the reactor contents. Switch off the stirrer.
8) Monitor the temperature. When the contents have cooled, open valve E to remove the product from the reactor.

When implemented by a computer, all of the above decision-making and timing is based upon software. For large chemical plants, such sequences can become very lengthy and intricate, especially when a plant involves many reaction stages. For the most efficient plant operation, a number of sequences might be in use simultaneously (e.g. in the context of the above example a number of reactions might be controlled at the same time). Very large process control computers are often dedicated almost exclusively to supervising such complex sequence-control procedures.

1.2.5 Supervisory Control Systems

Supervisory control systems are used to specify or optimise the operation of the set of DDC (or conventional analogue control) systems that are controlling a plant. For example, the objective of a supervisory system might be to minimise the energy consumption of a plant or to maximise its production efficiency. A supervisory system might compute the set points against which the plant control systems are to operate or it might re-organise the control systems in some way.

A simple example of where a supervisory control scheme can be utilised is illustrated in Fig. 1.4. Two evaporators are connected to operate in parallel. Material in solution is fed to each evaporator. The purpose of the plant is to evaporate as much water from the solution as possible. Steam is supplied to a heat exchanger linked to the first evaporator. Steam for the second evaporator is supplied from vapours boiled off at the first evaporation stage. To achieve maximum evaporation the pressures in the evaporation chambers must be as high as safety considerations will allow. However, it is necessary to achieve a balance between the two evaporators. For example, if the first evaporator is driven flat out then this might generate so much steam that the safety thresholds for the second evaporator are exceeded. A supervisory control scheme for this example will have the task of balancing the performance of both evaporators so that, overall, the most effective rate of evaporation is achieved.

In most industrial applications supervisory control, if used at all, is very simple and is based upon knowledge of the steady-state characteristics of the plant to define the required plant operating status. In a few situations, very sophisticated supervisory control algorithms have proved beneficial to plant profitability. Optimisation techniques using linear programming, gradient search methods (hill climbing), advanced statistics and simulation have been applied. In association with these techniques, complex non-linear models of plant dynamics and economics have been solved continuously in real time in parallel with plant operation, in order to determine and set up the most effective plant operating point. An example is the processing of crude oil by distillation. The most profitable balance of hydrocarbons can be produced under the direction of a complex supervisory control system.

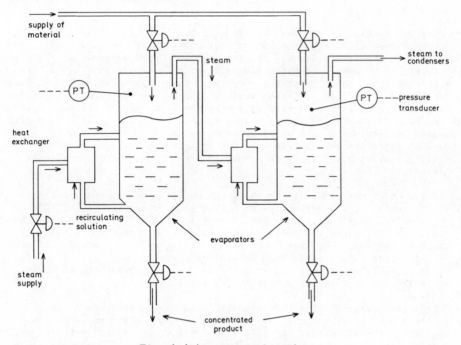

Fig. 1.4 An evaporation plant

1.2.6 The engineer and the process control computer

The control engineer has the task of specifying the various roles of the process-control computer and of implementing the specification. His duties may be itemised as follows (presuming the decision has been made as to the most suitable computer system for the job in hand):

1) To define measurements and actuations and set up scaling and filter constants, alarm and actuation limits, sampling intervals etc.
2) To define the DDC controllers, the interlinking/cascade connections between them and any other elements within the control system configuration.
3) To tune the above control systems, i.e. select appropriate gains, so that they perform according to some desired specification.
4) To define and programme the sequence control procedures necessary for the automation of plant operation.
5) To determine and implement satisfactory supervisory control schemes.

The control engineer may also have the job of determing how the plant operator is to use the computer system in the day-to-day running of the plant.

Clearly, for a large application, all of these duties would be beyond the scope of a single person and a team of people would be involved, one of whom would most certainly be a computer programmer. The latter phase, implementation of supervisory control, could be a project of many months if some of the more sophisticated techniques above mentioned are utilised.

The programming effort involved in establishing a complete working computer-control system could be considerable if the engineer had to start from scratch. However, process control applications have many aspects in common and standard packages of computer software are now available with many computing systems, thus minimising the effort required to establish a working control system. Facilities are available to permit the engineer to translate directly from specification charts and control system diagrams, such as that of Fig. 1.2, to a computer based implementation. The engineer defines the data base but does not have to write the software for the DDC systems. The standard software is usually sold with the computer system to form a complete process-control package.

There is a very significant short-coming with such standard packages. They provide for the definition and structuring of control systems but, to the author's knowledge, there is nothing currently available on the market that provides the control engineer with an aid for tuning the control systems. The engineer must rely completely upon his own experience of tuning three-term controllers. The quality of the tuning of industrial control systems is generally poor (perhaps because in lots of situations it does not matter much anyway).

Software packages for sequence control and supervisory control must be a lot more flexible than those for DDC. Sequencing and supervisory requirements will differ greatly from plant to plant whereas DDC configurations utilise a very limited range of standard operations. The standard packages therefore provide the engineer with a higher level language to programme the required sequence of commands. A variety of such languages exist. The most common are very similar to BASIC, with additional features to facilitate the real-time aspects and communications with the plant interface. If the language is interpretive (i.e. lines of programme are compiled at the time of execution), it is often possible to build up a sequence procedure while the computer is online to the plant. Supervisory control might require extensive calculations for which interpretive operation could be too slow and cumbersome. In this case, a language such as Fortran might be utilised and compilation would be necessarily offline. For the simpler supervisory schemes, the BASIC type languages are perfectly adequate.

1.2.7 Facilities for plant operator and plant manager

The plant operator must be provided with facilities that permit the straightforward operation of the plant on a day-to-day basis. The operator requires to be presented with all information relevant to the current state of operation of the controlled process and its control systems. In addition it is necessary for him to be able to interact with the plant, for example to change set points, to manually adjust actuators, to acknowledge alarm conditions etc.

A specially designed operator's control panel is a feature of nearly all computer control systems. Such a panel would typically consist of special keyboards, perhaps tailor-made for the particular plant that is controlled, and a number of display screens and printers. The video displays permit the operator to inspect, at various levels of detail, all monitored areas of the plant. The standard software packages supplied with computer control systems normally provide a range of display formats that can be used for the presentation of information. Typically, these might be an alarm overview display presenting information relating to the alarm status of large groups of measurements; a number of area displays presenting summaries of details concerning the control systems associated with particular areas of the process; and a large number of loop displays, each giving comprehensive details relating to a particular control loop. The control engineer selects the parameters that are to be associated with the individual displays, as part of the procedure of defining the data base for the computer system. The display presentations might be in the form of ordinary print-outs, or trend graphs of measurements, or schematic diagrams (mimics) of plant areas, with numerical data superimposed at appropriate locations. More expensive installations will utilise colour-graphic displays.

The above standard displays will not suit all requirements. For example, sequence control procedures might require special display presentation formats so that the operator can establish and interact with the current stage of process operation. Such displays would most likely be produced by programming them specially using the BASIC type language referred to above.

The special operator keyboards are usually built to match the standard display structures (for example by specifying particular keys to be associated with the selection of displays for particular plant areas). It is thereby straightforward for the operator to quickly centre upon aspects of interest. Commonly, if a particular control loop is pinpointed on a display, perhaps by use of a cursor or a light pen, then this will permit the operator to make direct adjustments to that loop using special purpose keys on a keypad.

The plant manager requires to access different information from the process control computer than the operator. He will need hard copy print-outs that provide day-to-day measures of plant performance and a permanent record of the plant operating history. Statistical analysis might be applied to the plant data prior to presentation to the manager so that the information is more concise and decisions are therefore more straightforward to make. The manager will be interested in assessing performance against economic targets, given the technical limitations of the plant operation. He will determine where improvements in plant operation might be possible. He will be concerned, along with the control engineer, with the operation of the supervisory control systems and will set the objectives for these top-level control systems.

1.3 HARDWARE FOR THE COMPUTER CONTROL SYSTEMS

In this chapter it has been assumed so far that the process control computer is a single hardware unit, that is, one computer performing all of the tasks itemised in section 1.2. This was nearly always the case with early computer control systems. The recent rapid developments of

solid state and digital technology have led to a very different approach
to the hardware configuration of computer control systems.

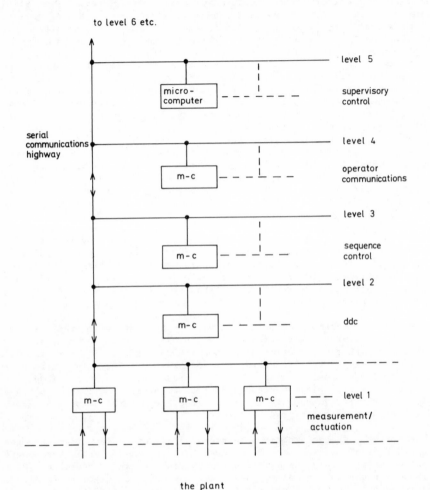

the plant

Fig. 1.5 A distributed and hierarchical microcomputer system

Fig. 1.5 illustrates a typical arrangement that might now apply with
many modern systems. The tasks of measurement, DDC, operator
communications and sequence control etc., are distributed amongst a
number of separate processing units, each of which will incorporate a
microcomputer of one form or another. The microcomputers are linked
together via a common serial communications highway and are configured
in a hierarchical command structure.

Fig. 1.5 indicates 5 broad levels of division in the structure of the
hierarchy of microcomputer units. These correspond in the main with
functions described in section 1.2.

Level 1:- all components and plant interfacing associated with measurement and actuation. This level provides the measurement and actuation data base for the whole system.
Level 2:- all DDC calculations.
Level 3:- all sequence control calculations.
Level 4:- operator communications.
Level 5:- supervisory control.
Level 6:- communications with other computer systems.

The boundaries between the levels do not have to be rigid, for example a unit for DDC might also perform some sequence control or might interface directly to the plant. The units for operator communications will drive the operator control panels and the associated video displays. The levels within the system define a command structure. Thus the microcomputer for supervisory control may direct the sequence control computers which, in turn, provide set points for the DDC computers etc.

The major features and advantages of this distributed/hierarchical approach are:

i) If the computational tasks are shared between processors, then system capability is greatly enhanced. The burden of computation for a single processor becomes very great if all of the described control features are included. For example, one of the main computing loads is that of measurement scanning, filtering and scaling. This is not because any one calculation is onerous but rather because of the large number of signals involved and the frequency at which calculations have to be repeated. Separation of this aspect from DDC, if only into two processors, greatly enhances the number of control loops that can be handled. The DDC computer will collect measurements already processed, via the communications link, at a much lower frequency than that at which the measurement computer operates.

ii) The system is much more flexible than a single processor. If more control loops are required or an extra operator station is needed, then all that is necessary is to add more boxes to the communications link. Of course, the units already in the system must be updated to 'be aware' of the additional items.

iii) If any unit should fail, the implications are not catastrophic, only one portion of the overall system will be out of commission, not the whole assembly.

iv) It is much easier to make software changes to the distributed system. For example, if a supervisory control programme is to be altered, then only the associated microcomputer need be called off-line. The risks of causing total system failure because of computer programming faults are very much reduced.

v) The units in the system can become standardised. This should lead to a much lower cost facility overall (some larger established manufacturers are still quoting prices that are of the same order as for older type computer control systems but smaller firms are beginning to dramatically undercut these prices). Thus, typically, a microprocessor for DDC might be standardised to cater for 16 loops. All of the necessary aspects such as gains, limits, set points, etc., would be

communicated from elsewhere via the highway. An application requiring 80
control loops would therefore utilise 5 such DDC boxes. In fact, at the
extreme low end of the computer control market there are microprocessor
units that implement a single control loop. These, to all external
appearances, apart from their data link facility, look very similar to
conventional electronic three-term controllers.

vi) The interlinking of the microcomputer units by a serial highway
means that they can be dispersed over quite a wide area. The highway may
easily stretch for a mile or even more if telecommunications devices are
used. An advantage is that it makes it unnecessary to bring many cables
carrying transducer signals to the control rooms. The measurement
microprocessors can be sited close to the source of the signals (i.e.
near the plant) and then only a serial link need be taken back to the
control room. Another advantage is that it is straightforward to set up
multiple operator control terminals at different sites in the factory.
Hierarchical distribution of process-control computers is not an
innovation with microcomputers. Very large chemical processes, or even
complete factories, for which a large capital outlay for control systems
can be justified, have employed the above principles. It is the
reductions in cost and size of the computing facilities that have made
the technology the most attractive for the majority of industrial
control applications today.

1.4 INTEGRITY OF COMPUTER CONTROL SYSTEMS

One of the main barriers to the use of computer control systems, apart
from expense, has been mistrust of the computer. A conventional
instrumentation and control system uses many individual units and the
failure of one or a few can be tolerated without having to shut down
plant operations. If all of these units are replaced by a single
computer then a break down of the computer can result in a complete loss
of all control systems, with unfortunate consequences. The typical
mean-time-between-failures for early computer control systems has been
in the region of 3 to 6 months. Many applications have therefore only
used a computer for aspects such as sequence and supervisory control for
which certain degrees of failure can be tolerated. Continuous feedback
control has, in these cases, continued to be implemented by electrical
or pneumatic instrumentation.

The solution to the problem of computer failure has been to provide
back-up systems to take over if a computer failure occurs. The backup
system might be a bare minimum of analogue controllers that are switched
in automatically. The difficulty with this approach is that if the
computer does not fail for a long time the plant staff might forget how
to operate the back-up system Wise users occasionally switch off the
computer deliberately. An alternative back-up mechanism is to duplicate
the computer system so that if one fails, another takes over. Certain
applications, e.g. in the nuclear industry, triplicate the control
computers. The above options are expensive and it has proved difficult
to establish change-over mechanisms that are guaranteed not to disrupt
the plant in any circumstances.

The new microcomputer technology has alleviated many of the problems
of integrity. Units, such as operator stations, may be duplicated on the
data highway at low cost. Units are now programmed to have a

self-diagnosis capability so that they can automatically establish and report if faults occur. The most vulnerable aspect of the new approach is the communications link, if this is broken, then all means of access to units on the wrong side of the break are lost, and for this reason, the communications link is often duplicated. Change over between links is automatic if a fault is detected. Some manufacturer's systems permit duplication to be extended to cover almost every unit within the distributed network. Mean-time-between-failures that would result in a total loss of plant control, are quoted at between 50 and 100 years. Needless to say, at such levels of duplication the systems cease to be cheap.

1.5 THE ECONOMICS OF COMPUTER CONTROL SYSTEMS

Before the advent of microprocessors, computer control systems were very expensive. When a new industrial process was being designed and built, or an old one being re-instrumented, a strong case had to be presented to use a computer in favour of conventional instrumentation.

In some cases computers have been used because, otherwise, plants could not have been made to work profitably. This is particularly the case with large industrial processes that require the application of complex sequencing procedures. The computer system permits repeatability in quality that is essential, for example, with manufacturing plants in the pharmaceutical industry. Flexibility of the computer is also important in these circumstances. Conventional instrumentation is difficult to modify if, for example, a sequencing procedure is to be altered to suit the manufacture of a different product. Reprogramming using a sequence language is comparatively straightforward.

Many large continuous processes (oil refineries, evaporators etc.) are also computer controlled. These processes stay in a steady state of operation for long periods and require little sequencing. They do incorporate many control loops. The usual justification for computer control in these circumstances has been that it will make plant operation more profitable. Such statements are often based upon the argument that even a small increase in production (say 1 to 2%) will more than pay for the computer systems. In the event, it has often been difficult to demonstrate that such improvement has resulted. The author is aware of one major installation where production has declined following the introduction of computers. The pro-computer lobby argue that it would have declined even more if the computer had not been there! The major benefits with continuous processes should arise through a better understanding of the process that invariably follows the application of computer control, and through the implementation of appropriate supervisory control schemes that maintain the plant closer to desired thresholds of operation.

The scene relating to the justification for computer control systems has now changed dramatically. The fall in cost and the improvement in reliabilty mean that for many industrial applications, both major and minor, it is automatic to install computer controls. In any event, microprocessor units are now cheaper than many of the equivalent analogue instruments, which will become obsolete before too long.

1.6 FUTURE TRENDS

The advent of the microcomputer has probably had more impact upon the discipline of control engineering than any other. Applications are now blossoming in all areas of industry on plants both large and small. The hardware revolution is still taking place but future changes are not likely to be as dramatic as in recent years. The question for the future is: 'Will there be an accompanying revolution in the control techniques that are implemented by the new hardware?'

An immense amount of effort has been devoted to academic control engineering research over the past two decades. Numerous novel control procedures have been postulated and proved effective, but only in numerical simulations. So many of the developed procedures could be summarised by the statement 'and here is yet another example of a controller for the linear plant described by the state of space equation $\dot{\underline{x}} = \underline{A} \underline{x} + \underline{B} \underline{u}$'. There has been very little application of these novel techniques, particularly in the process industries. They have remained very much the domain of the expert in applied mathematics. In industry the three-term controller has continued to be the fundamental control unit and will remain so for a long time to come.

The emphasis in control systems research and development must, and will, change. Much greater emphasis will be given to applying the new techniques and making them work. Computers will be used not only for the implementation of controllers but also to assist with defining controller configurations and with tuning, so that the best controlled performance is achieved. Techniques such as computer-aided design, identification of plant dynamics and adaptive/self-tuning control will become features of standard software packages supplied by the manufacturers of the hardware. It is possible to conceive of a new kind of box appearing on the data highway of Fig. 1.5, one for the design of control systems, a facility to assist the control engineer.

The increased emphasis on applications in control engineering research will motivate more problem specific solutions. The preoccupation of recent years with general theory will fade (as it must do anyway since linear control theory must be almost played out by now and extension to non-linear situations almost always requires a special case consideration). Greater attention will be given to consideration of supervisory control methods. This is the area where real financial returns from control systems can be made, by improving the productivity and efficiency of plant operations.

1.7 CONCLUSIONS

This chapter provides an introduction to the current state of the art of computer control systems. Many manufacturers provide standard system packages that implement the described features. Well known names such as Honeywell, Foxboro, Taylor and Kent cater for the distributed and hierarchical approach described in section 1.3 (TDC 2000, FOXNET, MOD III and P4000 respectively). Firms such as Negretti and Zambra, and Bristol Automation are marketing attractive microprocessor based systems (MPC 80 and MICRO B) for the smaller applications of say 50 control loops plus sequencing. Even the microprocessor companies themselves, such as Texas Instruments and DEC, are beginning to produce their own variations of distributed process control systems. The market is lively and competitive and costs are falling in line with the general downward trend of the price of computing equipment. Computer control is becoming commonplace.

Applications of new concepts in control theory have lagged far behind the boom in hardware. It is anticipated that this will change and that the technology of process control will be refreshed, in its turn, in the next few years. Facilities will become available to assist the control engineer to obtain better performance from the controlled plant than is currently possible unless a great deal of design effort is applied. The use of more sophisticated procedures for improved process control will become more common.

1.8 BIBLIOGRAPHIC NOTES

1. A best appreciation of the current state of computer systems is obtained by reference to manufacturer's publicity literature. Honeywell have produced a number of pamphlets under the title 'An Evolutionary Look at Process Control' in support of their TDC 2000 system. These provide a well presented overview of many of the aspects discussed in this paper. Taylor, Foxboro and Kent produce similar documentation relating to their new process control systems, but of rather less value educationally. Interesting contrasts between the various systems can be established by perusing this literature. A good sample of the capabilities of smaller systems can be established from the literature of Negretti and Zambra in support of their MPC 80 system.

2. For DDC:
   WILLIAMS, T.J.: 'Direct digital control and its implications for chemical process control', Dechema Monographien, 1965, v.53, No.912-924, pp.9-43
   DAVIES, W.D.T.: 'Control algorithms for DDC', Instrument Practice for Process Control and Automation, 1967. v.21, pp.70-77

3. For supervisory control concepts:
   LEE, GAINS, ADAMS.: 'Computer process control: Modelling and Optimisation', (Wiley, 1968)

4. For discrete control systems
   FRANKLIN, G.F., POWELL, J.D.: 'Digital Control of Dynamic Systems', (Addison Wesley, Reading,Mass,1980).
   CADZOW, J.A., MARTINS, H.R.: 'Discrete time and computer control systems' (Prentice Hall, Englewood Cliffs, 1970).

Chapter 2

Process control by computer

2.1 INTRODUCTION

This chapter is concerned with the control of slow-acting continuous processes typical of the chemical industries and also found in, for example, steam-raising for power generation, continuous bulk materials-handling and raw materials-processing. The variables to be regulated are usually the temperatures, pressures, levels and compositions of flowing fluids or pulverised solids and control exercised through the manipulation of flow-control valves, vibrating feeder devices etc. Such processes, before the advent of the on-line computer, were controlled by large numbers of standard measuring instruments individually connected to two- and three-term controllers, batteries of which were assembled into large control panels at some central control station. Indeed this remains the dominant practice today. The controller is usually of at least the indicating-type, displaying the current values of the set-point (reference), process-variable (feedback signal) and actuator-demand (controller-output, process-input) signals. In many instances, dedicated chart recorders are also incorporated within the instruments or associated with them for performance monitoring.

2.1.2 Evolution of computer control

Because of the large scale of the process industries, a thriving process instrumentation industry has grown up in the U.S.A. and the U.K., which supplies complete integrated instrumentation and control schemes with a high degree of standardisation. Each instrument and controller is readily adaptable to a wide range of process dynamics through the incorporation of widely adjustable parameter settings. Furthermore, instruments supplied are (a) ruggedised to withstand the arduous operating conditions e.g. extremes of temperature, humidity, vibration and dirt frequently encountered on plant, and perhaps more importantly, (b) intrinsically-safe. Intrinsic safety means that even under fault conditions e.g. accidental short- or open-circuit, any sparking produced in the internal or interconnecting circuitry of the instruments is incapable of igniting any surrounding inflameable vapour. In the mining and petro-chemical industries, such design features are clearly essential.

Because of the well established control practices already in operation therefore, the changeover to computer control in the process industries has been one of evolution rather than revolution. Existing methods of control have been progressively adapted to include the role of the computer. This situation contrasts with, say, machine control and particularly the numerical control of machine tools where digital control systems tend to be all-digital using, say, shaft-encoders or pulse-generators for measurement and stepping motors for actuation. In these applications complete analogue control systems, if used at all, were never available 'off-the-shelf' as in the process industries and furthermore digital measuring techniques in these applications provided a level of precision, often needed, but never readily obtainable using analogue techniques.

2.1.3 Supervisory control

In process control, analogue transducers were already capable of the necessary accuracy of measurement needed in this field so that initially the chief benefit provided by the process control computer was its ability to slowly update the set-points of the local analogue controllers in direct response to periodic recomputations of optimum process outputs (product yields, purity etc) according to varying operating conditions (selling price, energy costs etc). Before the advent of the on-line computer this task was performed by the process operator acting on data calculated off-line and hence the term Supervisory control: the computer merely supervising the control settings in the same way as the operator had hitherto supervised the process manually. With this form of control therefore the original analogue control loops are left virtually intact but the three-term controllers are fitted with motor-driven set-point potentiometers to permit their manipulation by the supervisory computer.

2.1.4 Direct digital control

After only a short period, however, the benefits of including the computer within the control loops, and indeed of interconnecting loops within the computer to achieve diagonal dominance, became more fully appreciated. This step of course converted the control loops to sampled-data systems and this form of control became termed Direct digital control (DDC). The computer could now act as a measurement filter (sampling-rate and noise-bandwidth permitting: see Shannon's sampling theorem), a diagonalising pre-compensator, as well as exercising one, two and three-term control action plus other dynamic shaping functions. Analogue instruments, because of their proven reliability achieved over many years of development, were retained as were the analogue three-term controllers. The analogue controllers were modified to act as standby controllers in the event of computer or data-link failure, for use during periods of plant commissioning prior to the computer coming on-line, during software modification and, in many instances, during plant start-up where the skills of a human operator may still be needed. (Recall that most process models, upon which theoretical control laws are based, are usually valid only for comparatively small perturbations about a steady state operating point. Such automatic control strategies are therefore only valid for process regulation around that point and not generally appropriate for running up a process from cold.) Standby analogue controllers are generally termed DDC back-up stations and, in addition to their emergency standby function, also provide a convenient means of interfacing a plant to a remote computer.

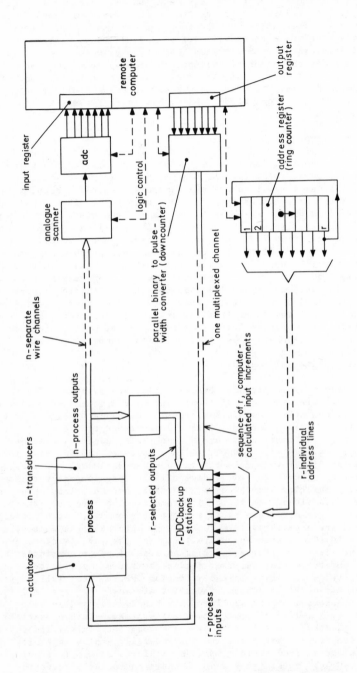

Fig. 2.1 Scheme for process control by remote computer

2.2 <u>TYPICAL CONTROL SCHEMES: HARDWARE FEATURES</u>

Fig. 2.1 illustrates schematically a type of multiloop, central computer control system frequently encountered nowadays in the process industries and will serve as a basis for discussion. Numerous variations in detail are of course potentially possible and actually operating and some of these latter variants are discussed in the following section.

2.2.1 <u>Data input methods</u>

Data may be transmitted from process to computer in a wide variety of forms, digital or analogue. Pulse-generating transducers, for instance, may have their pulse trains transmitted serially directly to the remote computer which may be programmed to count the pulses over fixed intervals of duration T (T >> mean pulse spacing) to determine the mean pulse-rate over each interval and hence estimate the mean value of the process variable over the same interval. Such data transmission is clearly of the digital type.

Alternatively, the transducer pulses, shaped to constant height and width may be fed into an analogue first-order lag filter of time-constant τ (again τ >> the mean pulse spacing) at either the process or computer end of the data link. In the former case the instrument output is effectively converted from a digital to an analogue type and data transmission is analogue in type. (Pulse generating transducers include, turbine flow-meters and radiation type instruments which measure either the thickness or density of a passing material by producing at a detector a proportional frequency of transmitted or backscattered radiation pulses from a radioactive isotope. Obviously some measurement lag is involved with this type of instrument because of the need for the counting interval T or smoothing time-constant τ). Digital transmission has the advantage that the presence or absence of a pulse is readily detected at the receiving end despite a substantial level of superimposed transmission noise that would seriously corrupt an analogue signal of similar strength. Furthermore, if noise conditions are severe, the reliability of digital data can be improved by the use of parity checking or data retransmission schemes. The disadvantage of transmitting the digital data direct is that valuable central processor (CPU) time is involved in pulse counting and parity checking unless separate peripheral digital equipment is provided at the computer interface.

For the moment, mainly because the majority of process transducers are essentially analogue devices, analogue data transmission from process to computer seems to hold sway as illustrated in Fig. 2.1, and the noise problem is overcome by transmitting high signal levels (typically 10-50 mA flowing in loops of 300 Ω resistance). This of course involves expensive data-transmission cables of considerable cross-section and schemes for remote data sampling and digitisation [1,2] are gradually becoming established, the sampled signals being transmitted serially along a single channel in pulse coded form i.e. the remote analogue instruments appear to the computer interface as a single sampled shaft-encoder.

Fig. 2.1 shows the analogue scanner situated at the computer interface so that the analogue measurements are transmitted individually and simultaneously (i.e. in parallel) to the computer interface. Siting the scanner at the plant would convert the data transmission to a pulse height modulation system (i.e. semi digital) and clearly would permit a reduction in the number of signal cable cores employed but involving greater complexity in the actual logic control of the scanner by the computer. Cable runs between instruments and any common gathering point are inevitably of long length in any case because most process plants are widely dispersed geographically. The relative potential saving in cable cost is therefore not necessarily substantial.

2.2.2 Data output methods

In transmitting data from computer to plant, as indicated in Fig. 2.1, a form of modulated pulse transmission is adopted although the information is conveyed not in the heights of the transmitted pulses but in their widths. The reason for the use of varying pulse area in output data transmission is that the desired incremental change in the selected process input is readily effected by merely applying the pulse to the input of a simple analogue integrator existing in each DDC backup station. Between pulses i.e. between samples, the integrator merely holds the process input constant as required. (The DDC backup station is considered in more detail in Section 2.2.4. They are typically designed to produce a 20% output swing for a pulse width of 25 m.s.). Varying pulse-widths or pulse-heights could of course produce a proportional change in integrator output equally effectively but pulse-widths tend to be preferred because of the ease with which the parallel digital signal produced by the computer may be converted to a proportional pulse width.

As indicated in Fig. 2.1, conversion is effected by means of a simple downcounter. This comprises a digital register into which the computer-calculated increments in each process input are loaded sequentially. After loading, the register counts down to zero and during the counting interval a fixed level is applied to the multiplexed data line, i.e. a fixed height pulse is transmitted having a duration proportional to the desired increment. (Alternatively the computer may itself effect the counting operation but, as with the counting of input pulses discussed above, this utilises valuable CPU time).

Control of the destination of each output pulse is essential to ensure that each desired process input increment is fed to the correct process actuator. For this purpose logic address lines are employed, one per actuator, running from the computer interface to the DDC stations as shown. The lines are raised (by a shift register in Fig. 2.1) sequentially and logic gating ensures that only the station addressed responds to the incoming pulse. The computer steps the address register between each outputting operation. Alternatively, of course, the address register could be located at the plant but the scheme shown clearly has the higher integrity. Since the address lines carry only digital signals of low signal strength, slender cables can be employed for the station addressing operation.

2.2.3 Interface control

The station addressing operation is but one example of the fact that not only plant data must pass across the computer/plant interface but that logic control data must also be transferred. This data is indicated

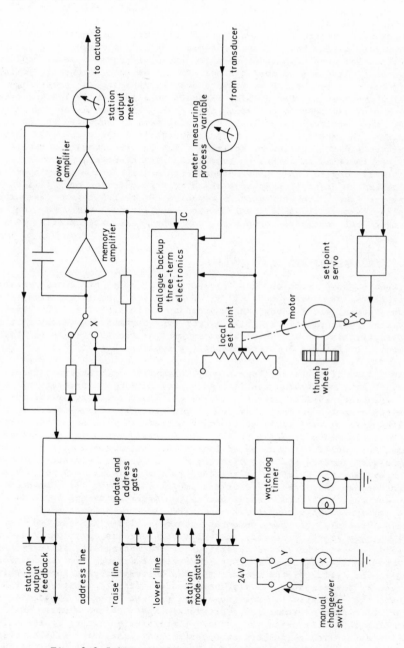

Fig. 2.2 Schematic diagram of DDC backup station

by the broken lines in Fig. 2.1. Where the computer is controlling its interface devices it is generally preferable that the computer not only instructs a peripheral device to operate (e.g. the analogue to digital converter to commence conversion, or the downcounter to begin the countdown) by the raising of the appropriate logic output line, but that the software instruction sequence should not recommence until the peripheral device has returned a logic input signal signifying that its operation has been completed. This technique of command and reply is known as <u>handshaking</u>. Since such operation times are frequently variable as in the case of the two examples above, handshaking is preferable to merely programming fixed delay segments to allow for device operation. Handshaking can therefore lead to time-saving in input/output operations where the devices have long operating periods compared to the basic computer instruction time. More importantly, however, it permits accurate fault location by software since the programme can readily be designed to raise an alarm, or take more drastic emergency action, if a peripheral device fails to respond within its known maximum operating period and to indicate precisely which interface device has failed. In this way faults are more quickly remedied. Of course in some very rapid sampling systems the additional logic input/output time may be ill afforded but chemical processes and the like are rarely in this category.

2.2.4 <u>Emergency backup</u>

Although some mention has already been made of local analogue controllers providing control in emergencies and other exceptional circumstances, and although their presence is indicated in the overall schematic diagram of Fig. 2.1, deeper consideration is necessary for a proper appreciation of the problems as well as the advantages of backup control. Further problems are exposed in sections 2.4 and 2.5.

Fig. 2.2 illustrates, again only schematically, the basic functions provided by a DDC backup station. As expected from inspection of Fig. 2.1 and the earlier general discussion, lines arriving at the station from the computer include a unique logic address line together with 'raise' and 'lower' lines multiplexed between all stations and carrying the variable width updating pulses. It is usual to have the separate lines for pulses to raise or lower the station outputs rather than a single wire implied at first sight by Fig. 2.1. The computer software arranges for the pulse to be switched to the appropriate line according to whether an increase or a decrease in station output is demanded by the control algorithm. Additionally, however, a unique logic line is also returned to the computer to indicate whether the station is on local control or on computer control. This is the so called <u>station mode status</u> line and, as will be seen in section 2.4, can vitally affect the correct operation of the controlling software. Also returned to the computer is an optional line carrying (as an analogue level) the value of the present station output (process input), which, if read by the computer via its analogue to digital converter (ADC), allows the software to know the actual process input as opposed to the value calculated by summing all previous output pulse widths. Use of this line thus eliminates the effect of calibration errors in the entire pulse transmission chain between computer and actuator. Alternatively, if the station manipulated, say, a flow-control valve, the feedback could be taken instead from a valve stem potentiometer or, better still, from a flow transducer so eliminating calibration errors in the process actuator system. All data flow between a station and the computer on the multiplexed lines is of course gated by the station's address line to ensure the correct routing of data around the numerous process control loops.

It is, of course, essential that the appropriate process output signal, generally termed the <u>process-variable</u> be monitored by the DDC station as indicated in Fig. 2.2 so that closed-loop control may be exercised by the station when required. The connection serves the other important function of ensuring that the local controller is properly <u>primed</u> for takeover of control at all times whilst the computer has control of process. To this end the local set-point potentiometer is motor-driven by a small servo-mechanism which, via relay contact X, nullifies the error between local set-point and process-variable while ever the station is on computer control. This covers the possibility of the set-point having been left manually at a dangerous value following, say, some special test on the process. The method assumes that, at the instant prior to changeover to local control, the computer control was already in a substantially steady-state so that the process variable should then equal the computer's internal set-point. Thus the motorised <u>tracking</u> by the local set-point of the process variable ensures the equality of the two set-points at changeover.

Obviously complete equality will not occur if changeover takes place during a transient disturbance of the process and this contingency could only be covered by transmission of the computer set-points for tracking by the associated back-up stations, necessitating a considerable increase in data-transfer.

The matching (or near matching) of set-points is only one of the requirements for smooth changeover of control, or <u>bumpless transfer</u> to use the conventional term. Bumpless transfer also demands the proper initialisation of all controller and software integrators and this topic is considered in Section 2.4.

Changeover to local control may usually be effected manually at the DDC back-up station by operation of the manual switch indicated in Fig. 2.2 which activates relay X thus stopping the set-point tracking and switching the memory amplifier input from the gated incoming pulse lines to the internal electronic closed-loop control network. (It will be noticed that the memory amplifier is converted from an integrator to a straight amplifier on changeover). Additionally changeover relay X may be energised by operation of relay Y driven from a <u>watchdog timer</u>. This timer is regularly reset by the arrival of successive address pulses at the station when under normal computer control but times-out, thus energising relay Y, if a pulse fails to arrive before the elapse of a preset period (> sampling interval). Failure of the data link, computer output interface or deliberate suppression of the address pulses by the computer program will therefore automatically cause changeover to local control. Return to computer control may be automatic on restoration of the address pulses or may necessitate a manual resetting of relay Y, both options usually being provided.

2.3 <u>SOFTWARE</u>

No attempt is made here to review the languages specially developed for real-time control e.g. CORAL [3], RTL/2 [4], MODULA [5] etc. Instead only the functional requirements for driving the scheme of Figs. 2.1 and 2.2 are briefly considered. Furthermore we neglect for the moment the software implications of having standby controllers present. These implications only become serious if dynamic control algorithms, i.e. those involving integrations, are employed and in this Section control laws will be assumed to be purely proportional, involving no dynamic compensation whatsoever.

The basic software operations required for controlling r loops all sampled at the same rate are shown in Fig. 2.3: the process inputs denoted by u_j, $0 < j <= r$ and the measured outputs by y_i, $0 < i <= n > r$. Measured process inputs are also assumed to be read-back into the computer via the analogue scanner (to avoid calibration errors as discussed in Section 2.2.4) and are distinguished from their computer calculated values by additional process output symbols i.e. $y_{n+j} \approx u_j$, $0 < j <= r$.

No programme initialisation is shown in Fig. 2.3 and only the repetitive process control segment is illustrated. This segment remains in a compiled form within the computer while ever control is in progress and is 'turned-on', i.e. made to run, usually at regular intervals: the sampling intervals. The turn-on operation may be effected within the computer itself which enters a delay loop between the completion of one programme execution and the elapse of the sampling interval or by triggering from an external interrupt signal produced by, say, a hardware timer. This may be preferable in situations where the computer handles a number of time-shared programmes having different levels of priority to ensure that routine operations such as data-logging should not hold up the execution of the control instructions when required. Certain occasional emergency control action e.g. in the event of an externally detected over-pressure may be needed at an even higher priority level than the repetitively executed continuous control law and a top priority interrupt would be allocated to turn on such an emergency routine.

Fig. 2.3 shows the basic software functions required to drive the system of Figs. 2.1 and 2.2. It shows: (a) the reading of the n process output measurements plus the r process input measurements, (b) the control algorithm for calculating the next desired values for the r process inputs, (c) the calculation of the increments Δu_i, $0 < i <= r$, required to be applied by the process actuators, (d) the outputting of these increments to the down-counter, (e) the stepping and raising of the station address lines and (f) the handshaking operations between computer and interface devices.

Obviously many variations of this basic software scheme are conceivable and possible and many additional features could be included. One essential additional feature is necessary once dynamic control algorithms are included and this problem is considered in Section 2.4.

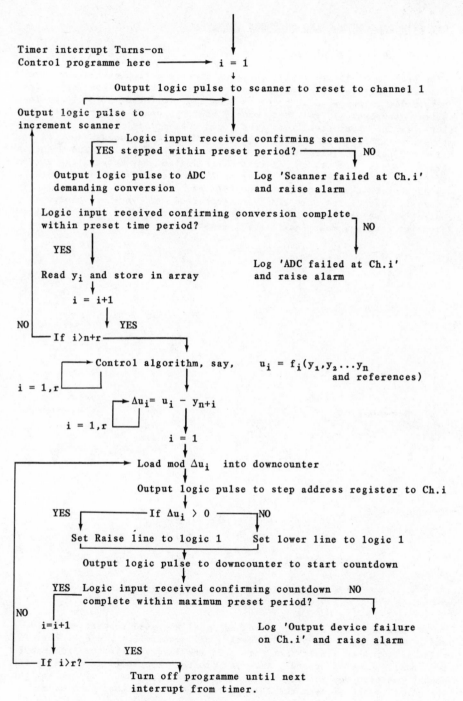

Fig. 2.3 Basic software flowchart for driving system of Fig. 2.1

2.4. INTEGRAL CONTROL AND BUMPLESS TRANSFER

2.4.1 At the back-up station

The problem of incorporating integral control action within a hardware two- or three-term controller has been well understood for many years. The problem arises when a process loop is switched off closed-loop control temporarily and onto manual control. Under such circumstances there is no guarantee that the controller set-point and the process-variable will remain in equality since the process input is now being manipulated independently of the controller's demand. Unless special steps are taken in controller design, any integrating action will tend to 'wind-up' to its upper or lower saturation limit over a period under these open-loop conditions. (The same problem clearly occurs if control is temporarily transferred to another closed-loop controller unless the two are tuned precisely identically and their set-point settings made equal at all times).

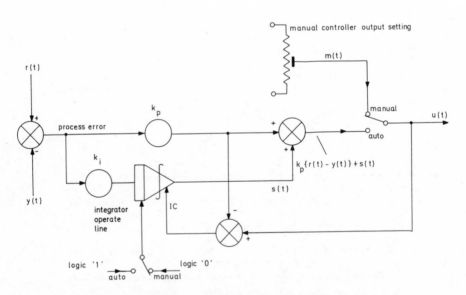

Fig. 2.4 Scheme for integrator initialisation in a
hardware two-term controller

The solution is straightforward and is illustrated schematically for a two-term proportional-plus-integral (PI) controller in Fig. 2.4. The circuit indicates simply that, if t_0 is the time of transfer from manual to local automatic control, then the process input $u(t)$ is given by the normal two-term control algorithm after transfer, i.e. if r is the set-point, y the process variable and k_p the proportional gain then:

$$u(t) = k_p\{r(t) - y(t)\} + s(t), \quad t \Rightarrow t_0 \qquad (2.1)$$

where the integral action

$$s(t) = k_i \int_{t_o}^{t} \{r(t) - y(t)\}dt, \quad t => t_o \tag{2.2}$$

where k_i is the integral gain. Prior to transfer, however, $u(t)$ is equated to the manual setting $m(t)$ so that

$$u(t) = m(t), \quad t < t_o \tag{2.3}$$

and the integral action initialised such that

$$s(t) = m(t) - k_p \{r(t) - y(t)\}, \quad t < t_o \tag{2.4}$$

Hence

$$u(t_o) = k_p \{r(t_o) - y(t_o)\} + s(t_o) \tag{2.5}$$

and, if t_o- denotes a time marginally before transfer

$$s(t_o-) = m(t_o-) - k_p \{r(t_o-) - y(t_o-)\} \tag{2.6}$$

Now if $m(t)$ and $r(t)$ are held constant over the interval $t_o- < t <= t_o$, since y is the output of a continuous dynamic process its value too will be unchanged over the instant of transfer so that under these circumstances $s(t_o-) = s(t_o)$ and by adding eqn. (2.5) and (2.6) we get

$$u(t_o) = m(t_o-) \tag{2.7}$$

The process input therefore remains unchanged across the instant of transfer from manual to automatic control whenever this occurs provided the integral action is continuously primed beforehand according to eqn. (2.4).

The foregoing applies also to situations where the process input may be provided temporarily by other controllers, rather than merely a manual setting, provided $m(t)$ is now taken to mean the 'process input from whatever the alternative source'. This source, in a DDC situation, would of course include <u>the value applied by the computer</u> when the local station is switched to <u>computer control</u>.

2.4.2 <u>Software for bumpless transfer</u>

An important lesson to be learned from the foregoing consideration of local hardware controllers is that software controllers too must continue to monitor the plant and the action of any standby controllers even when the software is itself not exercising full control of the plant. This is to allow the software to take over control bumplessly at any time. It is therefore essential for the control programme to be running <u>before</u> control is transferred to the computer and for the software to know, (a) whether or not it has control of the plant and (b) if not, the process input values applied to the plant from elsewhere i.e. from the standby controllers (whether these be acting in a local closed- or open-loop manner). There is, therefore, an additional need for process actuator signals to be fed back to the computer over and above that discussed in Sections 2.2 and 2.3.

Fig. 2.5 outlines a software segment for the PI control of a single loop of the process assuming that the output y_i is interlinked with input u_i. Obviously a diagonally dominant process may be treated as merely a multiplicity of such systems and each loop treated independently i.e. some computer loops may be switched in and some out quite freely. Interactive processes need rather more care, however, if integrity is to be retained and this problem is considered in Section 2.5.

Fig. 2.5 demonstrates the importance of the station mode-status line in informing the computer whether or not it has control of the loop in question and the right-hand branch of the flow-chart shows the initialisation of the integral action according to eqn. (2.4). The left-hand branch shows merely a discrete equivalent of PI control using Simpson's rule for calculation of the error integral.

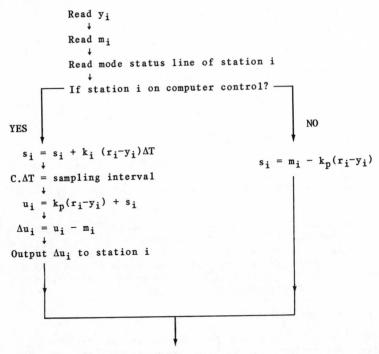

Read y_i

Read m_i

Read mode status line of station i

If station i on computer control?

YES

$s_i = s_i + k_i (r_i - y_i)\Delta T$

C.ΔT = sampling interval

$u_i = k_p(r_i - y_i) + s_i$

$\Delta u_i = u_i - m_i$

Output Δu_i to station i

NO

$s_i = m_i - k_p(r_i - y_i)$

Fig. 2.5 Software segment for bumpless-transfer of a single PI control loop

2.4.3 General

The foregoing remarks concern one example only and are intended to serve only as an indication of the way in which proper priming of dynamic control laws must be exercised whenever changes to and from computer control are necessary. This problem clearly becomes crucial in cases of so called 'high-integrity computer control systems' where standby computers may be employed in addition to the main computer. As already mentioned however, the standby or backup concept can lead to additional integrity problems in the case of highly interactive multivariable processes. These are considered in the next section.

2.5. INTEGRITY OF INTERACTIVE SYSTEMS

For our purposes, a control system which remains stable on the loss of one or more actuator or measurement signals is said to have __integrity__.

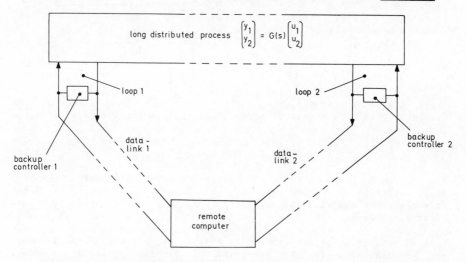

Fig. 2.6 Remote computer system for a distributed 2x2 system

2.5.1 Theoretical background

We shall confine attention to 2x2 systems in the interests of simplicity although the findings of this Section may be readily generalised to multi-input multi-output systems. Consider the long, physically-distributed process illustrated in Fig. 2.6. The process is actuated at opposite ends by inputs u_1 and u_2 as indicated and process variables y_1 and y_2 measured in the vicinity of u_1 and u_2 respectively. The process may be controlled by a central remote computer over data links 1 and 2 as indicated but, because of their long length and consequent vulnerability to failure, standby controllers 1 and 2 would normally be provided interlinking u_1 to y_1 and u_2 to y_2 respectively by compact, local, emergency control loops 1 and 2, again as shown in Fig. 2.6.

Regarding y_1, y_2, u_1 and u_2 as small deviations from the steady-state outputs and inputs of the process then the process may normally be described by a linear transfer function model of the form:

$$\begin{bmatrix} y_1 \\ y_2 \end{bmatrix} = \underline{G}(s) \begin{bmatrix} u_1 \\ u_2 \end{bmatrix} \tag{2.8}$$

where transfer function matrix (TFM) $\underline{G}(s)$ is given by

$$\underline{G}(s) = \begin{bmatrix} g_{11}(s) & , & g_{12}(s) \\ g_{21}(s) & , & g_{22}(s) \end{bmatrix} \qquad (2.9)$$

i.e. $\qquad\qquad y_1 = g_{11} u_1 + g_{12} u_2 \qquad\qquad\qquad (2.10)$
and $\qquad\qquad y_2 = g_{21} u_1 + g_{22} u_2 \qquad\qquad\qquad (2.11)$

Now while ever <u>both</u> data links 1 and 2 are healthy, the computer may exercise a partial diagonalising function by implementing, in the software, a control law of the type, say,

$$\begin{bmatrix} u_1 \\ u_2 \end{bmatrix} = - \underline{K}(s)\ \underline{Q}(s) \begin{bmatrix} y_1 \\ y_2 \end{bmatrix} \qquad (2.12)$$

where $\underline{K}(s)$ is a diagonal gain matrix or perhaps a diagonal matrix of PI elements, $\underline{Q}(s)$ is a matrix of simple digital lead/lag elements designed such that

$$\underline{Q}(j\omega_1) \simeq \underline{G}^{-1}(j\omega_1) \qquad (2.13)$$

and ω_1 some fixed angular frequency chosen to give nearly complete diagonalisation of the process over the frequency range of interest (i.e. around the critical -1 point). With well behaved systems, ω_1 may often be taken as zero.

In situations where interaction is slight, the precompensator $\underline{G}^{-1}(j\omega_1)$ may be discarded in the interests of simplicity leaving a purely diagonal controller and a small degree of interaction to be tolerated on closed-loop control. On backup control, however, in the event of failure of say, data-link 1, control of y_1 would fall back to local loop 1 whilst y_2 could either continue on computer control (under a modified algorithm) or be allowed by the computer to fall back to local control by loop 2. In either event diagonal control is now the only option in the absence of an additiional data link between the local controllers. (This would be equally vulnerable to failure as those between the plant and the computer in the case of a widely distributed plant of the type considered here). We, therefore, now consider the effect of diagonal control on a non-diagonal process.

We consider first the case of <u>loop 1 closed</u> with <u>loop 2 open</u>. From eqn. 2.10, if loop 2 is open, $u_2 = 0$ so that the open-loop transfer-function of loop 1 is given by

$$y_1/u_1 \Big|_{u_2 = 0} = g_{11} \qquad (2.14)$$

Now consider the hypothetical case of <u>loop 2 on perfect closed-loop control</u>, i.e. $y_2 = 0$. We now have, from (2.11) that

$$0 = g_{21}u_1 + g_{22}u_2$$

so that

$$u_2 = -(g_{21}/g_{22})u_1 \qquad (2.15)$$

and using eqn. (2.15) to eliminate u_2 from (2.10) we find that with loop

2 perfectly closed, the open-loop transfer function for loop 1 becomes

$$y_1/u_1 \Big|_{y_2 = 0} = g_{11} - g_{21}g_{12}/g_{22}$$

$$= g_{11}(1 - g_{12}g_{21}/(g_{11}g_{22})) \qquad (2.16)$$

The gain of loop 1 therefore changes on perfect closure of loop 2 from g_{11} by an increment $-g_{21}g_{12}/g_{22}$, this change of course being frequency dependent in general. Clearly, provided

$$|(g_{12}g_{21})/(g_{11}g_{22})| \ll 1.0 \quad , \quad \text{all } \omega \qquad (2.17)$$

the interaction problem is slight but if condition (2.17) is not satisfied serious disturbance to the tuning of loop 1 would be caused by the subsequent closure of loop 2. In fact, closure (or opening) of loop 2 could bring about the complete destabilisation of loop 1 if this were originally tuned with loop 2 open (or closed). Since neither the loop 2 controller nor the computer knows the state of loop 1 (i.e. whether it is on closed-loop or manual control) once data link 1 has failed, the only safe action is to ensure that one or other of the standby controllers always falls back to manual control in the event of loss of central computer control whenever a process fails to satisfy condition (2.17). Local backup of central computer control systems must therefore be applied with some caution.

Very often the location of actuation and/or measuring points is based on a somewhat arbitrary decision at the plant design stage. A flow-control valve, for instance, may often be sited at either end of a pipe-line. It may therefore be possible to overcome the integrity problem by correct physical sitings and this ensures that condition (2.17) is satisfied at the outset. The control engineer should therefore involve himself closely with the plant design rather than merely waiting to attempt to compensate for bad plant design subsequently at the control system design stage. An example of this is considered in the next section.

2.5.2 Distillation column example

Fig. 2.7 shows a binary distillation column diagrammatically. Columns are generally very tall plants exposed to the outside environment and there is, therefore, great incentive for keeping local control loops compact. The five dependent variables to be controlled are, usually, the liquid levels H_a and H_b in the accumulator and reboiler at the top and bottom of the column respectively, the pressure P of vapour within the column, and, most importantly, the compositions X_a and X_b of the top and bottom products produced by the column. Measurements of X_a and X_b are normally inferred from measurements of boiling points Y_a and Y_b made at two points near the top and bottom of the process respectively as indicated.

Five manipulated variables are available for exercising the necessary control, these being V, the vapour rate, C the condenser coolant rate, D the top product or distillate rate, W the bottom product rate and R the rate at which liquid is refluxed (re-cycled) back into the column. These variables are all indicated on Fig. 2.7 and a problem clearly exists in choosing the best pairing of dependent and independent variables (i.e. in choosing which output should manipulate which input).

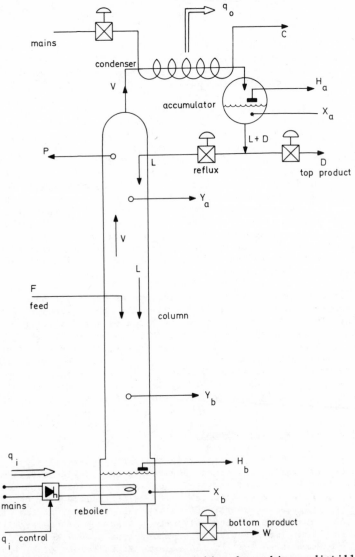

Fig. 2.7 Manipulated and control variables for a binary distillation column (L above equivalent to R in text)

The major dynamic problem concerns the control of X_a and X_b and much can be done to ease this problem by making the best choice of two variables from V, C, D, W and R to exercise this control. This choice should be made to minimise interaction effects whilst leaving local control loops as compact as possible.

The pressure variations are governed by

$$dP/dt = k_1(q_i - q_0 - \text{heat losses}) \qquad (2.18)$$

where k_1 is approximately constant and q_i and q_0 are the rates of heat supplied to the reboiler and removed via the condenser respectively.

Now
$$V = q_i/\tilde{L} \qquad (2.19)$$
and
$$C = q_0/k_2 \qquad (2.20)$$

where k_2 is nearly constant and $\tilde{L}$ is the latent heat of the distilled mixture (again almost constant) so that column pressure, which may be measured at top or bottom, may be controlled by manipulation of either V (via q_i) or C so that the back-up pressure control loop may be sited at the top or bottom of the process.

Considering now the level control we have that, under constant pressure

$$dH_a/dt = k_a(V - R - D) \qquad (2.21)$$
and
$$dH_b/dt = k_b(R + F - V - W) \qquad (2.22)$$

where k_a and k_b are constants so that H_a may be controlled by manipulation of V, or preferably for a physically tight control loop, by manipulation of R or D. Bottom level H_b may clearly be controlled by R or preferably by V or W.

The important conclusion to be drawn from the above analysis is that, whilst retaining compact pressure and level control loops, flows at either top or bottom of the process remain for composition control.

We shall consider first control by V and D for which the process model (for a fairly symmetrical plant), in terms of small perturbations x_a, x_b, v and d may be approximately expressed as follows if tight level and pressure control are assumed:

$$\begin{bmatrix} x_a \\ x_b \end{bmatrix} = g(s) \begin{bmatrix} \varepsilon & -1 \\ -\varepsilon & -1 \end{bmatrix} \begin{bmatrix} v \\ d \end{bmatrix} \qquad (2.23)$$

where $g(s)$ is a scalar transfer function which is closely approximated by a first-order lag. (Lower case symbols are here used to denote small perturbations)

Interlinking x_a with v and x_b with d
<u>Interlinking x_a with v and x_b with d</u>

Using our general notation of Section 2.5.1 this coupling implies setting $y_1 = x_a$, $u_1 = v$, $y_2 = x_b$ and $u_2 = d$ giving an interaction term

$$\frac{g_{12}\,g_{21}}{g_{11}\,g_{22}} = \frac{(-\varepsilon)(-1)}{\varepsilon(-1)} = -1 \qquad (2.24)$$

The interaction term is therefore of significant magnitude but, because its sign is negative (see equation 2.16) will tend to merely cause a doubling of loop gain on closure of a second loop. Tuning is, therefore, affected but destabilisation need not occur if a suitably large gain margin is allowed. The interlinking is not ideal practically however since, although V may be effectively manipulated at the top of the column (i.e. near X_a) by adjustment of C (if P sets q_i) the D-valve, as shown, is remote from the X_b measurement.

Interlinking x_a with d and x_b with v

We again have that $y_1 = x_a$, $y_2 = x_b$ but now $u_1 = d$ and $u_2 = v$ giving

$$\frac{g_{12}\ g_{21}}{g_{11}\ g_{22}} = \frac{\varepsilon(-1)}{(-\varepsilon)(-1)} = -1 \tag{2.25}$$

Similar interaction effects to the foregoing case are therefore produced by this arrangement which is practically superior since X_a and D are physically close as are X_b and V (if P now sets C).

Interlinking x_a with v and x_b with r

Assuming tight level control at all times then the feedrate F into the column must always balance the total take-off rate so that

$$D + W = F \tag{2.26}$$

Similarly considering each end vessel separately, again with tight level control, it follows that

$$V = R + D \tag{2.27}$$

and

$$F + R = V + W \tag{2.28}$$

Clearly (2.27) and (2.28) are consistent with equation (2.26) and from these equations we deduce that for constant F, small perturbations v, r, d and w in the flow rates are related thus

$$d = -w \tag{2.29}$$

and

$$v = r + d = r - w \tag{2.30}$$

These flow equations allow the composition dynamics (equation 2.23) to be investigated when inputs other than v and d are used. We here consider inputs v and r.

Now from (2.30) it follows that

$$\begin{bmatrix} v \\ d \end{bmatrix} = \begin{bmatrix} 1 & 0 \\ 1 & -1 \end{bmatrix} \begin{bmatrix} v \\ r \end{bmatrix} \tag{2.31}$$

so that in terms of v and r the composition model becomes

$$\begin{bmatrix} x_a \\ x_b \end{bmatrix} = g(s) \begin{bmatrix} \varepsilon & -1 \\ -\varepsilon & -1 \end{bmatrix} \begin{bmatrix} 1 & 0 \\ 1 & -1 \end{bmatrix} \begin{bmatrix} v \\ r \end{bmatrix} \tag{2.32}$$

$$= g(s) \begin{bmatrix} \varepsilon-1 & 1 \\ -(\varepsilon+1) & 1 \end{bmatrix} \begin{bmatrix} v \\ r \end{bmatrix}$$

Choosing $y_1 = x_a$, $y_2 = x_b$, $u_1 = v$ and $u_2 = r$ therefore giving

$$\frac{g_{12}\ g_{21}}{g_{11}\ g_{22}} = \frac{-(\varepsilon+1)\ 1}{(\varepsilon-1)\ 1} = \frac{1+\varepsilon}{1-\varepsilon} \simeq 1+2\varepsilon \tag{2.33}$$

since

$$0 < \varepsilon << 1.0 \tag{2.34}$$

The interaction term is therefore marginally greater than unity and being positive causes a complete reversal of loop 1 gain sign on closure of loop 2. The gain clearly changes from g_{11} to $-2\varepsilon g_{11}$ (see equation 2.16). This coupling of inputs and outputs is clearly totally unsatisfactory.

Interlinking x_a with r and x_b with v
_____

With this reserve coupling (which from Fig. 2.7 is clearly superior from practical siting considerations) the interaction term now becomes

$$\frac{g_{12}\,g_{21}}{g_{11}\,g_{22}} = \frac{(\varepsilon-1).\,1}{-(\varepsilon+1).\,1} = \frac{1-\varepsilon}{1+\varepsilon} \simeq 1-2\varepsilon \qquad (2.35)$$

The loop 1 gain therefore changes from g_{11} to $+2\varepsilon g_{11}$ on closure of loop 2 so that the sign of the gain is retained but its value drastically altered, necessitating the use of an impracticably large gain margin. Other interlinkings could clearly be explored and are left as an exercise for the reader.

2.6 CONCLUSION

In this chapter some of the problems of applying remote computer control to chemical plants and similar processes have been examined concentrating paricularly on the provision of back-up control for high system integrity. Emphasis has been placed on the practical aspects of computer control system design. The more theoretical and purely computational aspects are covered in the chapters which follow.

2.7 REFERENCES

1. Barnes, R.C.M.: 'CAMAC – a review and status report', Proc. IEE International Conference on 'Trends in On-line Computer Control Systems', IEE Conf. Pub. No. 127, April 1975, p.15.
2. Halsall, J.R., and Kirby, I.J.: 'MEDIA: A continuous digital process control system', Proc. IEE International Conference on 'Trends in On-line Computer Control Systems', IEE Conf. Pub. No. 85, April 1972, p.185.
3. Neve, N.J.F.: 'CORAL 66 – The U.K. national military standard', Proc. of IEE International Conference on 'Trends in On-line Computer Control Systems', IEE Conf. Pub. No. 127, April 1975, p.139.
4. Benson, R.S.: 'Experience with a MEDIA/RTL-2 process control system', ibid, p.1.
5. Wirth, N.: 'Modula: a language for modular programming', Software-practice and Experience, 1977, v.7, pp.3-35; Wirth, N.: 'The use of Modula', Software-Practice and Experience, 1977, v.7, pp.37-65.

Chapter 3

Some DDC system design procedures

3.1 INTRODUCTION

As a result of the enormous impact of microprocessors, electronic engineers, with sometimes only a cursory background in control theory, are being involved in direct digital control (DDC) system design. There is now a real need for an easily understood and simply implemented design technique for single-input DDC systems which is the objective of this chapter. The proposed design procedure is shown as a flow chart in Fig. 3.1, and it contains appropriate references to sections of text that treat the more important issues in detail. It is hoped that this diagram will provide a systematic approach to DDC system design, as well as an appreciation for the organisation of the text. The experienced designer will notice the absence of such topics as:

(i) 'Bumpless transition' criteria during changes from automatic to manual control.
(ii) Provision for timed sequential operations to cope with fault conditions.

These aspects have been deliberately omitted because they are considered to be separate programming issues once control in the normal operating regime has been achieved.

Digital realisations of conventional three-term controllers have the advantage of: wide applicability, theoretical simplicity and ease of on-line tuning. However, the resulting closed-loop performance is generally inferior to that obtainable with other algorithms of similar numerical complexity. The graphical compensation procedure [2] described in Section 3.3 copes with the design of digital three-term controllers and these high performance units with equal ease. Also the technique is readily exploited by virtue of its simple calculations (amenable even to slide-rule treatment). Furthermore, it is shown to result in an 'even-tempered' closed-loop response for all input signals; unlike DDC systems synthesised by time-domain procedures.

Compared to analogue controllers, digital controllers offer distinct advantages in terms of: data transmission, interconnection, auxiliary data processing capabilities, fault tolerance, tamper resistance etc. However, a digital controller must evidently provide a control performance at least as good as that of the analogue controller it replaces. In this respect, it is suggested that sampling and wordlength effects are designed to be normally negligible relative to the control accuracy specification. When this objective is frustrated by computer performance constraints these degenerate effects can be evaluated from formulae given in the text.

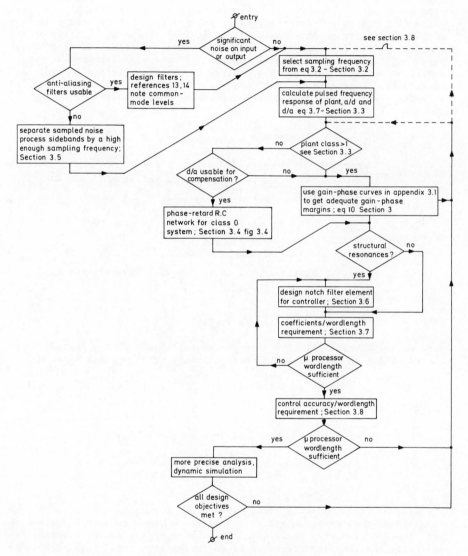

Fig. 3.1 DDC design scheme

3.2 CHOICE OF SAMPLING FREQUENCY

The first step in design of a DDC system of the form shown in Fig. 3.2 is the selection of an appropriate sampling rate (T). Distortion in the form of spectral side-bands centred on integral multiples of the

sampling frequency (1/T) is inherently produced by the periodic sampling
of information, and the perfect (impractical) recovery of the signal
requires the complete elimination of these, harmonics [3]. A suitable
practical choice of sampling frequency limits the distortion (or
aliasing) by imposing a large enough frequency separation between the
side-bands and the original unsampled signal spectrum for the low-pass
plant elements to effect an adequate attenuation. Where comprehensive
plant records for an existing analogue control scheme are available, the
sampling period for a replacement DDC system is sometimes decided on the
basis of requiring a 'small change' in the time dependent error or
deviation signal during this interval. In the author's opinion,
engineering judgements of this form are valuable only as confirmation of
a fundamentally based and experimentaly confirmed analysis.

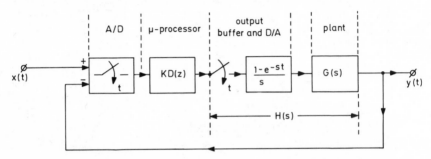

Fig. 3.2 Single input, unity feedback DDC system

For a sinusoidal input:

$$x(t) = A \sin(wt) \tag{3.1}$$

Knowles and Edwards [4] show that the average power of the control error
component due to imperfect side-band attenuation is bounded by:

$$\overline{\varepsilon^2_R} \;<= \; A^2 g^2 w_s^{-(2R + 2)} M_p^2 w^2 \; |G(jw)|^{-2} \tag{3.2}$$

where:

$$G(s) \simeq \frac{g}{s^R} \Bigg|_{s \to \infty} \; ; \; M_p = \text{Closed-loop peak magnification} \tag{3.3}$$

$$w_s = 2\pi/T$$

Equation (3.2) reveals that the high frequency components of the
continuous input spectrum generate relatively the greatest distortion
and loss of control accuracy. Thus a simple design criterion for
selecting the sampling frequency is to ensure that the right-hand-side
of equation (3.2) represents an admissible loss of control accuracy for
the largest permissible amplitude sinusoid at the highest likely signal
frequency. This calculation is not only easily performed, but it is
independent of subsequent stabilisation calculations. However, if
computer loading demands the lowest possible consistent sampling
frequency, it is then necessary to follow an interactive procedure which
involves the digitally compensated system and the formula [4]:

$$\overline{\varepsilon^2}_R = \pi^{-3} g^2 w_s^{-2} R \int_{-\infty}^{\infty} |K^*(jw)/H^*(jw)|^2 \sin^2(wT/2) \, \emptyset_x(w) \, dw \qquad (3.4)$$

where $K^*(jw)$ defines the overall pulsed frequency response of the stabilised closed-loop system, and $\emptyset_x(w)$ is the input signal power spectrum.

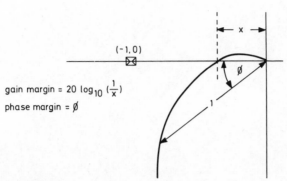

Fig. 3.3 Nyquist diagram illustrating the gain/phase margin criterion

3.3 FREQUENCY DOMAIN COMPENSATION METHOD

For most practical applications, necessary and sufficient conditions for the DDC system in Fig. 3.2 to be stable are that the open-loop pulsed frequency response:

$$KD(z)H(z)\Big|_{z \,=\, \exp(j2\pi fT)} \;\overset{\Delta}{=}\; KD^*(jf)H^*(jf) \qquad (3.5)$$

does not encircle the $(-1, j0)$ point. Furthermore, given that the polar diagram of $KD^*(jf)H^*(jf)$ is of the general form shown in Fig. 3.3, then an adequately damped time domain response is produced when the defined Gain and Phase Margins are greater than about 12dB and 50° respectively. It is evidently important to confirm that these stability margins are maintained over full-range of plant parameter variations or uncertainties. Denoting the Laplace transfer function of the plant, ADC, DAC and data smoothing elements by $H(s)$ then the corresponding pulsed frequency response is obtained [3, 5] as:

$$H^*(jf) = \frac{1}{T} \sum_{-\infty}^{\infty} H(j2\pi f - j2\pi n f_s) \qquad (3.6)$$

Data sampling frequencies selected according to the criterion in Section 3.2 allow the above expression to be closely approximated by:

$$H^*(jf) \simeq \frac{1}{T} H(s)\Big|_{s \,=\, j2\pi f} \qquad (3.7)$$

over the range of frequencies that determine closed-loop system stability. If the data smoothing element consists of a zero-order hold, and the combined gain of the ADC and DAC converters is K_T, the required pulsed frequency response $H^*(jf)$ is readily calculable from:

$$H^*(jf) = K_T \exp(-j\pi fT) \cdot \left[\frac{\sin(\pi fT)}{\pi fT} \right] \cdot G(s)\Big|_{s=j2\pi f} \qquad (3.8)$$

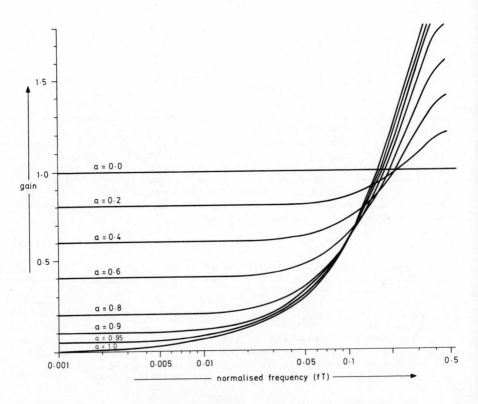

Fig. 3.4 Graph of the magnitude of (exp(j2πfT)−a) to base fT

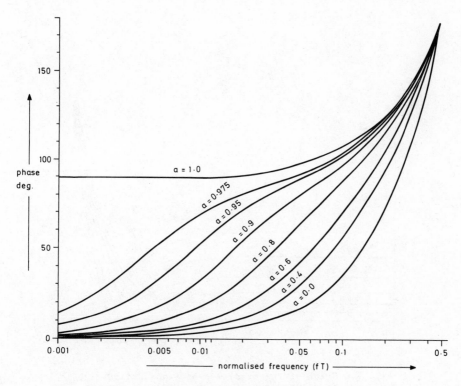

Fig. 3.5 Graph of the phase of (exp(j2πfT)−a) to base fT

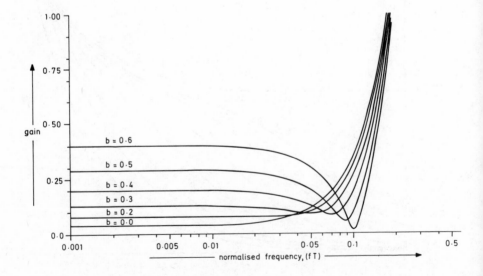

Fig. 3.6 Graph of the magnitude of $(\exp(j2\pi fT-0.8-jb)$
$(\exp(j2\pi fT-0.8+jb)$ to base fT

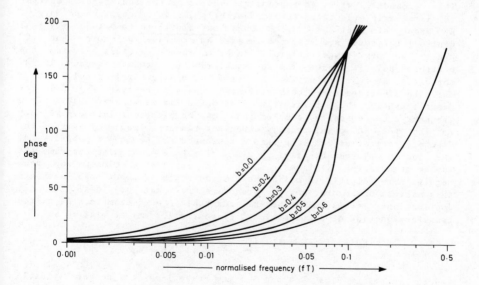

Fig. 3.7 Graph of the phase of (exp(j2πfT−0.8−jb))
        (exp(j2πfT)−0.8+jb) to base fT

For ease of nomenclature in the following discussion, the number of pure integrations in the plant transfer function G(s) is termed the 'class' of the system. In general, it is recommended that one stage of phase-lead compensation is included for each pure integration in the plant. Thus for a class 1 system, the discrete compensator takes the form:

$$KD(z) = K \left(\frac{z - a_1}{z - a_2}\right) \text{ with } 0 <= a_2 < a_1 < 1 \qquad (3.9)$$

where the pole at $z = a_2$ is necessary for physical realisability. Observe that digital realisations of PI and PD controllers can be represented in the form of equation (3.9). Typical graphs of the gain and phase of $[\exp(j2\pi fT) - \alpha]$ to base of normalised frequency (fT) are given in Figs. 3.4, 3.5, 3.6 and 3.7 for real and complex values of α, which corresponds to a controller zero or pole location. As with analogue networks, increasing amounts of phase-lead are seen to be associated with increasingly severe rising gain-frequency characteristics. If the maximum phase-lead per stage is much above the recommended value of 55°, the rising gain-frequency response of the discrete controller causes the locus KD*(jf)H*(jf) to bulge close to the (-1, j0) point which aggravates transient oscillations.

Pursuing a maximum bandwidth control system design, the measured or calculated values of H*(jf) are first used to determine the frequency (f_b) for which:

$$\underline{/H*(jF_b)} + 55° \text{ x Number of Phase-lead Stages} = -130° \qquad (3.10)$$

(a Bode or Nyquist diagram is helpful in this respect)

This procedure with an appropriate choice of the scalar gain constant (K) is clearly orientated towards establishing an adequate Phase Margin. By means of Fig. 3.5, which shows $\underline{/(\exp(j2\pi fT) -\alpha}$ to base of fT, real controller zero(s) are selected to give on average no more than 70° phase-lead per stage at $f_B T$. In the case of a class 1 system, the controler pole is chosen from the same graph to produce around 15-20° phase-lag at $f_B T$. For a class 2 system, complex controller poles have a definite advantage over their real counterparts, because Fig. 3.7 for example shows that relatively less phase-lag is generated at a given frequency. Consequently, complex poles may be placed relatively closer to the (-1, j0) point thereby enhancing the zero-frequency gain of the discrete controller, and the load disturbance rejection capability of the control system. Finally, by means of a Bode or Nyquist diagram, the scalar gain factor (K) is set so as to achieve a closed-loop peak magnification of less than 1.3 and a Gain-Margin of about 12dB. Although this maximum bandwidth design procedure may not be ideal for all applications it is nevertheless adaptable. For example, the pulsed transfer function of a conventional PI controller amy be written as:

$$KD(z) = (K_P + K_I) \left(\frac{z - a}{z - 1}\right) ; \quad a = \frac{K_P}{K_P + K_I}$$

and the graphs in Figs. 3.4 and 3.5 may be applied to select suitable values for 'a' and K_P and K_I, which uniquely specify the controller. Again, the design curves may be used to match the frequency response function of a digital controller to that of an existing analogue controller for the range of frequences affecting plant stabilisation.

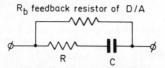

Fig. 3.8 Phase-retard compensation using DAC

3.4 THE COMPENSATION OF CLASS 0 (REGULATOR) SYSTEMS

The performance and stability of class 0 systems are generally improved by the inclusion of phase-lag compensators, whose analogue transfer function has the general form:

$$R(s) = K \left(\frac{1 + s\alpha\tau}{1 + s\tau}\right) \tag{3.11}$$

where the attenuation constant α is less than unity but greater than 1/12 for practical applications. An ADC [6] is frequently time-division multiplexed between several controllers, but each loop normally has exclusive use of its own DAC which can serve useful dual roles. Fig. 3.8 shows a series resistor and capacitor connected accross the feedback resistor (R_b) of a DAC, and this is proposed for the phase-lag compensation of class 0 DDC systems. Apart from the benefits accruing from savings in computing time, the use of analogue networks in this instance is superior to digital because experience shows that an appreciable phase-lag then occurs over a markedly narrower range of frequencies for the same attenuation constant. The importance of this fact will become apparent after the following description of phase-lag network design.

The first step in the compensation process is to determine the frequency w_B rads/sec for which the phase of the frequency response function $H*(jw)$ is $-130°$. If the Gain and Phase Margins of the uncompensated regulator are satisfactory, but the problem is to reduce the zero-frequency control error to ε_{DC}, then set

$$10/\alpha\tau = w_B \tag{3.12}$$

and

$$\alpha = \left(\frac{\varepsilon_{DC}}{1 - \varepsilon_{DC}}\right) H*(0)$$

$$K = 1/\alpha \tag{3.13a}$$

With these parameters the gain and phase of $R(s)$ for frequencies above w_B are effectively unity and zero respectively. Hence inclusion of the network increases the zero-frequency gain of the open-loop response without altering the original satisfactory stability margins. Alternatively, if the zero-frequency control accuracy is acceptable, but improvements in stability margins are desired, then select parameters according to eqn. (3.12) and

$$K = 1$$

$$\alpha = 1/|H*(jw_b)| \text{ for a } 50° \text{ Phase Margin} \tag{3.13b}$$

With these values, the gain and phase of R(s) for frequencies above w_b are α and zero respectively. Hence inclusion of the network improves the Gain Margin by the factor 20 $\log_{10}$ (1/α), without altering the original satisfactory open-loop zero-frequency gain. Both the above design techniques for phase-lag networks are based on sacrificing phase in a relatively unimportant portion of the Nyquist diagram for gain in a separate region that markedly influences control accuracy or stability. As the frequency range over which the network produces an appreciable phase-lag widens, this simple design procedure becomes evidently complicated by the progressive overlap of these regions of gain and phase changes.

3.5 NOISY INPUT OR OUTPUT SIGNALS

The operational amplifier associated with a DAC can also be used for additional signal conditioning. Sometimes the input or output signal of a control system is heavily contaminated by relatively wide-bandwidth noise. If it is impractical to low-pass filter these signals to prevent aliasing [13, 14), then an irrevocable loss of control performance can only be prevented by employing a sampling frequency which largely separates the side-bands of the sampled noise process. Unnecessary actuator operations, and therefore wear and tear, are sometimes avoided in this situation by including an averaging calculation in the digital compensator. In these circumstances, the side-band attenuation achieved by the plant is rendered less effective if the output sampling rate of the controller is reduced to an integral sub-multiple of its input rate. As an alternative solution to this noisy signal problem, it is proposed that a suitable capacitor is placed in parallel with the feedback resistor of the DAC.

3.6 STRUCTURAL RESONANCES AND DIGITAL NOTCH NETWORKS

A servo-motor is coupled to its load by inherently resilient steel drive shafts, and the combination possesses very selective resonances because the normal dissipative forces are engineered to be relatively small. In high power equipment (radio telescopes, gun mounting etc.) the frequency of such structural resonances can be less than three times the required servo-bandwidth, and consequently their presence markedly complicates the compensation procedure. Viewed on a Nyquist diagram, the resonances cause the locus to loop-out again from the origin to encircle the (-1, j0) point. Analogue controllers obtain system stability under these conditions by processing the error or deviation signal with tuned notch filters (eg Bridged-Tee) which block the excitation of each oscillatory mode. For a resonance peak at w_0 rad/sec, the same effect is provided by the following digital filter:

$$N(z) = K_N \frac{(z - z_0)(z - z_0^*)}{(z - r_0 z_0)(z - r_0 z_0^*)} \tag{3.14}$$

where:

$$z_0 = \exp(jw_0 T) \; ; \; z_0^* = \exp(-jw_0 T) \; ; \; 0 \leq r_0 < 1 \tag{3.15}$$

and unity gain at zero-frequency is achieved by setting:

$$K_N = \left| \frac{1 - r_0 z_0}{1 - z_0} \right|^2 \tag{3.16}$$

In order to null a mechanical resonance, its bandwidth and centre-frequency must both be matched by the digital filter. As may be appreciated from the pole-zero pattern of the filter, no larger notch bandwidth than necessary should be used in order to minimise the phase-lag incurred at servo-frequencies. It remains therefore to relate the notch bandwidth (B_{NO}) to the parameter r_0.

The gain-frequency response function of the proposed digital filter is given by:

$$|N^*(jw)| = K_N \left| \frac{(\exp(jwT) - z_0)(\exp(jwT) - z_0^*)}{(\exp(jwT) - r_0 z_0)(\exp(jwT) - r_0 z_0^*)} \right| \qquad (3.17)$$

Defining the frequency deviation variable:

$$\delta = w - w_0 \qquad (3.18)$$

then a power series expansion of exponential functions of δ yields the first order approximation:

$$|N^*(j\delta)| = \frac{K_{NO}}{\sqrt{1 + (\frac{1 - r_0}{T\delta})^2}} \quad \text{for } \delta T \ll 1 \qquad (3.19)$$

where:

$$K_{NO} = K_N \left| \frac{1 - \exp(-j2w_0 T)}{1 - r_0 \exp(-j2w_0 T)} \right| \qquad (3.20)$$

Equation (3.19) evidently has complete similarity with the gain-frequency characteristic of an analogue notch filter in the vicinity of its anti-resonance. Accordingly, its bandwidth is defined as the frequency increment about w_0 within which the attenuation is greater than 3dB:

$$B_{NO} = \pm (\frac{1 - r_0}{T}) \qquad (3.21)$$

Calculated or measured plant frequency responses may therefore be used in conjunction with equations (3.15) and (3.21) to specify suitable values for w_0 and r_0 in equation (3.14), and the resulting filter is incorporated as an algebraic factor in the pulse function of the discrete controller.

3.7 COEFFICIENT QUANTISATION IN A DISCRETE CONTROLLER

The coefficients in a pulse transfer function usually require rounding in order to be accommodated in the finite wordlength format of a microprocessor or minicomputer. Because generous stability margins are normally employed, coefficient rounding is normally unimportant as regards the practical realisation of phase-lead or phase-lag pulse transfer functions. However, it is well-known that increasingly severe digital filter specifications (cut-off rate, bandwidth etc.) accentuate computer wordlength requirements [8, 9]. Hence it is generally prudent to examine the wordlength necessary to counter acute structural resonances by the proposed form of digital notch filter.

By writing the eqn. (3.14) as:

$$N(z) = K_N \; \frac{z^2 - az + 1}{z^2 - rbz + r^2} \tag{3.22}$$

coefficient rounding on $\{a, \; rb, \; r^2\}$ is seen to modify both the centre-frequency and the bandwidth of a notch filter design. Ideally with no rounding the coefficients in eqn. (3.22) are specified by:

$$a = a_0; \; (rb) = (r_0 b_0); \; a_0 = b_0 = 2 \cos (w_0 T); \; (r^2) = (r_0^2) \tag{3.23}$$

and defining for the practical situation the variables

$$a = 2 \cos (w_1 T); \; b = 2 \cos (w_2 T) \tag{3.24}$$

$$\delta = w - w_1 \quad ; \; \varepsilon = w_1 - w_2 \tag{3.25}$$

One obtains the gain-frequency response function of the realisations as:

$$|N^*(jw)| = K_N \left| \frac{\exp(j\delta) - 1}{\exp(j\delta + j\varepsilon) - r} \right| \; \left| \frac{1 - \exp[-j(w + w_1)T]}{1 - r\exp[-j(w + w_2)T]} \right| \tag{3.26}$$

Thus the centre-frequency of the filter is not at w_1, and the shift due to the finite computer wordlength is evaluated directly from eqn. (3.24) as:

$$\delta w_0 = \frac{1}{T} \; [\cos^{-1}(a/2) - \cos^{-1}(a_0/2)] \; \text{rad/sec} \tag{3.27}$$

It is interesting to note that the change in centre-frequency cannot be obtained by differentiating eqn. (3.24) because a second-order approximation is required if the coefficient a_0 approaches -2. In the vicinity of the anti-resonance and with fine quantisation:

$$\delta T \ll 1 \; \text{and} \; \varepsilon T \ll 1 \tag{3.28}$$

and under these conditions a power series expansion of the exponential terms in eqn. (3.26) yields:

$$|N^*(j\delta)| \simeq \frac{K_N'}{\sqrt{(1 + \varepsilon/\delta)^2 + ((1 - r)/\delta T)^2}} \tag{3.29}$$

where:

$$K_N' = K_N \left| \frac{1 - \exp(-j2w_1 T)}{1 - r \exp(-j2w_2 T)} \right| \tag{3.30}$$

Thus the bandwidth (B_N) of the realisation satisfies:

$$(1 + \frac{\varepsilon}{B_N})^2 + (\frac{1 - r}{B_N T})^2 = 2 \tag{3.31}$$

or:

$$B_N^2 = 2\varepsilon B_N + \varepsilon^2 + (\frac{1 - r}{T})^2 \tag{3.32}$$

For small enough perturbations about the ideal situation defined in eqn. (3.23), it follows that:

$$\left|\delta B_{NO}\right| \, <= \, \left|\frac{\partial B_N}{\partial \varepsilon}\right|_O \, \left|\delta \varepsilon\right| \, + \, \left|\frac{\partial B_N}{\partial r}\right|_O \, \left|\delta r\right|$$ (3.33)

From eqn. (3.23), (3.25) and (3.32) one obtains:

$$\delta \varepsilon \, <= \, 2 \, \delta w_O; \quad \delta r \, <= \, q/2 r_O$$
$$\left|\frac{\partial B_N}{\partial \varepsilon}\right|_O \, = \, 1; \quad \left|\frac{\partial B_N}{\partial r}\right|_O \, = \, 1/T$$ (3.34)

where the width of quantisation is defined by:

$$q \, = \, 2^{-(\text{wordlength} \, - \, 1)}$$ (3.35)

Hence the change in the notch bandwidth of the filter due to coefficient rounding is bounded by:

$$\left|\delta B_{NO}\right| \, <= \, 2\left|\delta w_O\right| \, + \, \frac{q}{2T \, (1 \, - \, B_{NO} \, T)}$$ (3.36)

As a design example, consider the realisation of a notch filter with: centre frequency 12 rad/sec, bandwidth ± 1 rad/sec and a sampling period of 0.1 sec. Eqn. (3.23) defines the ideal numerator coefficient as:

$$a_O \, = \, 0.724716$$ (3.37)

With an 8-bit machine format, the actual numerator coefficient realised is:

$$a \, = \, 0.718750 \quad (0.1011100)$$ (3.38)

so that from eqn. (3.27):

$$\delta w_O \, = \, 0.032 \text{ rad/sec}$$ (3.39)

and the change in filter bandwidth evaluates from eqn. (3.36) as:

$$\delta B_{NO} \, <= \, 0.11 \text{ rad/sec}$$ (3.40)

Thus in practical terms, an 8-bit microprocessor is about sufficient for realising this particular filter specification.

3.8 ARITHMETIC ROUNDOFF-NOISE IN A DISCRETE CONTROLLER

Arithmetic multiplications implemented by a digital computer are subject to rounding errors due to its finite wordlength. As a result, random noise is generated within the closed-loop and the control accuracy is degenerated. In extreme cases, the control system can even become grossly unstable. It is therefore important to relate the loss of control accuracy to the wordlength of the discrete controller. The analysis in [10] provides an easily calculable upper-bound for the amplitude of the noise process present on the error signal due to arithmetic rounding in a discrete controller. With a directly programmed compensator for example, this upper-bound is given by:

$$\left|\varepsilon_Q\right| \, <= \, (\frac{\text{Number of Significant Multiplications in KD(z)}}{2K\left|\Sigma \text{ Numerator Coefficients}\right|}) \, q$$ (3.41)

where a 'significant' multiplication does <u>not</u> involve zero or a positive integral power of 2 including 2^O. During a normally iterative design

procedure, eqn. (3.41) is valuable as a means of comparing the
multiplicative rounding error generated by various possible controllers.
In preliminary calculations, the pessimism of this upper-bound is of
little consequence (the formula over-estimated by about 2-bits in the
example considered in [10]). However, a more accurate estimate of
performance degeneration due to multiplicative rounding errors can be
economically important for the finalised design when it is realised on a
bit-conscious microprocessor. For this purpose, the analysis in [11] and
[12] is recommended because one straight-forward calculation quantifies
the loss of control accuracy as a function of controller wordlength.

As described earlier, the sampling frequency of a DDC system must be
high enough to prevent an unacceptable loss of control accuracy or
fatigue damage to plant actuators. However, strangely enough, it is
possible in practical terms for the sampling frequency to be made too
high. Reference to Fig. 3.5 shows that by increasing the sampling
frequency, the phase-lead required at a particular frequency can only be
sustained by moving the compensator zero(s) closer to unity. To preserve
load disturbance rejection, the zero-frequency gain of the controller
$(KD(z)|_{z = 1})$ must be maintained by then moving the pole(s) also closer
to unity. As a result, the algebraic sum of the compensator's numerator
coefficients is reduced, and multiplicative round-off noise on the error
signal is increased according to Equation (3.41). Thus the choice of an
unnecessarily high sampling frequency can entail unjustified expense in
meeting long wordlength requirements.

3.9 MULTIRATE AND SUBRATE CONTROLLERS

A subrate digital controller has its output sampler operating an
integral number of times slower than its input. In a multi-rate
controller, the ouput sampler operates an integral number of times
faster than its input. Section 3.5 describes how subrate systems
prejudice ripple spectrum attenuation, and in practice excessive
actuator wear seems best prevented for noisy input signals by the use of
a single rate controller and a smoothing capacitor across the DAC unit.
It is now pertinent to question if multi-rate systems afford any
advantages over their more easily designed single-rate counter-parts.

As a result of the higher pulse rate exciting the plant, Kranc [15]
contends that the ripple performance of a multi-rate system is superior
to that of a single-rate system, even though both cases have the same
input sampling rate. However, it should be noted that the input sampler
of the multi-rate sampler still generates spectral side-bands centred on
multiples of $\pm w_o$, and that these are not eliminated in further
modulation by the faster output sampler. Hence it is by no means
self-evident that increasing the output sampling rate alone in a digital
compensator effects an improvement in ripple attenuation. To clarify
this issue and other aspects of performance, a comparison of single and
multi-rate unity feedback systems is implemented in [16] for the plant
and DAC transfer function

$$G(s) = \frac{e^{-sT}(1 - e^{-sT})}{s^3(s + 1)} \qquad (3.42)$$

with computer sampling periods of:

$$T_{in} = T_{out} = 0.1 \text{ sec} \quad \text{Single-rate system}$$

$$(3.43)$$

$$T_{in} = 3T_{out} = 0.1 \text{ sec} \quad \text{Multi-rate system}$$

Both closed-loop systems are compensated to have virtually identical frequency and transient response characteristics. For two typical input spectra, calculations show that the output ripple power for the multi-rate system is at least a decade greater than that for the single-rate system. An intuitive explanation of this result may be based on the fact that in both systems the compensator and zero-order hold are attempting to predict some function of the error over the input sampling period. As each prediction inevitably incurs error, a multi-rate system makes more errors per unit time and thus gives the larger ripple signal. It may be argued that multi-rate controllers give potentially the faster transient response. However, the conclusion reached in [16] is that improvement in transient response leads to a further degradation in ripple performance.

3.10 TIME DOMAIN SYNTHESIS WITH POLYNOMIAL INPUTS

After a finite number of sample periods, a dead-beat DDC system tracks a specified input test polynomial exactly with no intersample error. As the output of a zero-order hold remains constant during an intersample period, it follows that with this data smoothing unit a dead-beat system is obtained only if:

(i) The number of integrations in $G(s)$ is at least equal to the degree of the specified input polynomial;
(ii) the closed-loop pulse transfer function is constructed to be a finite polynomial in z^{-1} (i.e. all poles of $K(z)$ at the origin).

It follows from Fig. 3.2 that the pulse transfer function for the computer output sequence is $K(z)/H(z)$. Therefore condition (ii) for a dead-beat response is satisfied when $K(z^{-1})$ is set equal to any finite polynomial in z^{-1}, which includes all zeros and the transport delay of $H(z)$. Note that extra coefficients may be included in $K(z^{-1})$ to meet other design requirements. Having thus devised a closed-loop pulse transfer function to achieve a dead-beat response and other design criteria, the digital compensator required is given by:

$$D(z) = \frac{1}{H(z)} \frac{K(z)}{1 - K(z)}$$

$$(3.44)$$

As an example [3] of a dead-beat design, suppose a 1 second sampling period, a zero-order hold and the plant:

$$G(s) = \frac{10}{s(s+1)^2}$$

$$(3.45)$$

The z-transformation of the hold and plant is:

$$H(z) = \frac{z^{-1}(1 + 2.34 \ z^{-1})(1 + 0.16 \ z^{-1})}{(1 - z^{-1})(1 - 0.368 \ z^{-1})^2}$$

$$(3.46)$$

so that condition (ii) implies that:

$$K(z^{-1}) = z^{-1}(1 + 2.34 \ z^{-1})(1 + 0.16 \ z^{-1})(a_0 + a_1 z^{-1} + a \ z^{-1} + \ldots)$$

$$(3.47)$$

or,

$$1-K(z^{-1}) = 1 - a_0 z^{-1} - (a_1 + 2.5a_0)z^{-2} - (a_2 + 2.5a_1 + 0.3744a_0)z^{-3} -$$
$$(a_3 + 2.5a_2 + 0.3744a_0)z^{-4} + \text{etc.}$$

$$(3.48)$$

If the other design requirement is zero steady-state error for a ramp input, then as the control error sequence is:

$$E(z^{-1}) = [1 - K(z^{-1})] \frac{z^{-1}}{(1 - z^{-1})^2} \qquad (3.49)$$

it follows that $1 - K(z^{-1})$ must include $(1 - z^{-1})^2$ as a factor so that:

$$1 - K(z^{-1}) = (1 - z^{-1})^2(b_0 + b_1 z^{-1} + b_2 z^{-2} + \ldots) \qquad (3.50)$$

After equating coefficients of z^{-1}, it evolves that $\{a_0 \; a_1\}$ and $\{b_0 \; b_1 \; b_2\}$ are the only consistent and linearly independent sets to satisfy simultaneously eqns. (3.48) and (3.50) with

$$a_0 = 0.73; \; a_1 = -0.47$$

Thus the synthesised dead-beat closed loop pulse transfer function is:

$$K(z^{-1}) = 0.73 \; z^{-1} + 1.35 \; z^{-2} - 0.9 \; z^{-3} - 0.18 \; z^{-4} \qquad (3.51)$$

A minimum variance DDC system achieves zero steady-state control error at the sampling instants for a particular polynomial input, and it also minimises the sum of the squares of the control error sequence for the same or a different polynomial input. For example, if the design criterion is for zero steady-state control error for a ramp input, then eqn. (3.50) defines $1 - K(z^{-1})$ and the corresponding error sequence for step input is:

$$E(z^{-1}) = z^{-1} (1 - z^{-1})(1 + b_1 z^{-1} + b_2 z^{-2}) \qquad (3.52)$$

and as,

$$\sum_{n=0}^{\infty} e_n^2 = \frac{1}{2\pi j} \int_{|z| = 1} E(z) \; E(z^{-1}) \; z^{-1} dz \qquad (3.53)$$

then by Cauchy's Theorem:

$$\sum_{n=0}^{\infty} e_n^2 = 1 + (b_1 - 1)^2 + (b_2 - b_1)^2 + b_2^2 \qquad (3.54)$$

Differentiating the above variance yields the condition for its minimum as:

$$b_1 = 2/3, \; b_2 = 1/3 \quad \text{with} \quad \sum_{n=0}^{\infty} e_n^2 = 4/3 \qquad (3.55)$$

and,

$$K(z^{-1}) = 4/3 \; z^{-1} - 1/3 \; z^{-2} \qquad (3.56)$$

Attention could now be directed at the constraints in the above domain synthesis techniques; such as that all poles and zeros of $H(z)$ outside the unit circle should be included as factors of $1 - K(z)$ and $K(z)$

respectively (see Chapter 7, of [3] for example). However, it is more relevant to compare these time domain synthesis methods with the frequency domain compensation procedure described earlier in Section 3.3.

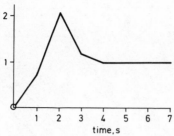

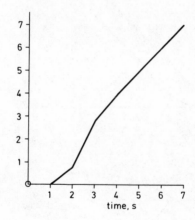

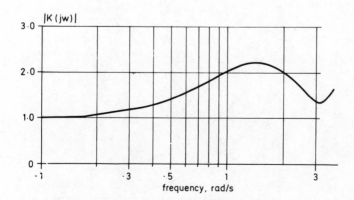

Fig. 3.9 Response characteristics of dead-beat system
a) step response b) ramp response c) closed-loop frequency

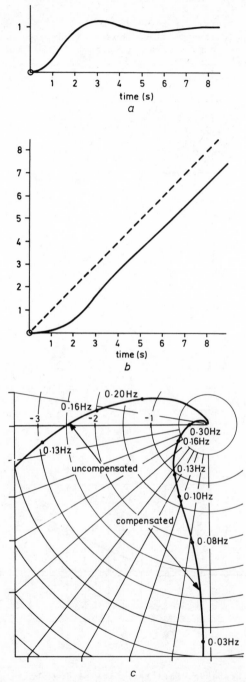

Fig. 3.10 Response characteristics of frequency domain compensated system
a) step response b) ramp response c) closed-loop frequency response

3.11 COMPARISON OF TIME DOMAIN SYNTHESIS AND FREQUENCY DOMAIN COMPENSATION TECHNIQUES

A convenient starting point is Fig. 3.9 which shows that the step response of the time domain synthesised dead-beat system has a 100% overshoot, even though its ramp response is acceptable. Insight into this behaviour may be gained from the closed-loop frequency response $K^*(jw)$ of the system, which is also shown in Fig. 3.9. Noting that the spectrum $X^*(jw)$ of the input sequence $\{x_n\}$ always completely defines the sequence and vice-versa, then as:

$$Y^*(jw) = K^*(jw) \, X^*(jw) \qquad (3.57)$$

it follows that minimal transmission distortion occurs when $K^*(jw)$ is unity over the range of significant spectral components. The overall frequency response in Fig. 3.9 displays a marked resonance at 1.6 rad/s, and the wider frequency spectrum of a step function compared to a ramp produces the relatively stronger excitation observed. For many purposes, the 100% overshoot in the step response would prove unacceptable.

With the proposed frequency domain compensation procedure, it is recalled that the Class 1 system in eqn. (3.45) is compensated by a series element of the form:

$$D(z) = K\frac{(z - a)}{(z - b)} \qquad (3.58)$$

whose maximum phase-lead must not exceed about 55°. In order to achieve an adequate phase-margin ($\simeq 50°$), the frequency (f_b) at which the phase lag of $H^*(jw)$ is 180° indicates the maximum bandwidth likely to be achieved. It is readily seen from eqn. (3.45) that in this case:

$$f_B \, T \simeq 0.16 \qquad (3.59)$$

However, Fig. 3.5 shows that the phase difference between the numerator and denominator of a discrete compensator is then well below the maximum, which occurs around:

$$fT = 0.03 \qquad (3.60)$$

More effective compensation can therefore be realised by decreasing the sampling period to 0.25 seconds (note the rms ripple error at 1 rad/s even with a sampling period of 1s is satisfactory; eqn. (3.2) evaluates as 1.3×10^{-3}). The pulse frequency response of the open-loop uncompensated system is given in Fig. 3.10, and calculations based on eqn. (3.8) differ in gain and phase by less than 0.01% from those computed using the z-transformation. As $f_B \, T$ now equals 0.035, a digital compensator which provides 55° lead at this frequency is selected by means of Fig. 3.5 as:

$$D(z) = K\frac{(z - 0.9)}{(z - 0.4)} \qquad (3.61)$$

and a gain constant (K) of 0.4 realises a gain-margin of approximately 10 dB. The step and ramp function responses of this frequency domain compensated system are also shown in Fig. 3.10, and the additional control error due to multiplicative rounding errors is derived from eqn. (3.41) as:

$$|\varepsilon_Q| \Rightarrow 37.5 \ q \qquad\qquad (3.62)$$

which indicates that double precision arithmetic would be necessary on an 8 bit μ-processor. Observe that the proposed design technique has produced:

   (i) a control system which is reasonably damped for both step and ramp inputs;
  (ii) a conclusion that the rms ripple error is negligible;
 (iii) a simpler, and therefore more easily 'tuned', digital compensator;
  (iv) quantification of the stability margins which are available to cope with plant parameter variations or uncertainties;
   (v) a conclusion that double precision arithmetic is necessary if the compensator is to be realised on an 8 bit micro-computer.

Finally, it should be noted that a control system's response must be judged in general terms of achieving a small enough error with actual plant inputs. Frequently, step or ramp inputs cannot be considered as representative of the real situation. Under such conditions, these simple test inputs form only a rapid method of discovering resonances in the closed-loop frequency response, which would cause distortion in the transmission of actual input signals.

3.12 CONCLUSIONS

A complete design scheme for single-input direct digital control systems has been presented. This includes a simple calculation for ensuring that the sampling rate is consistent with a system's accuracy specification or the fatigue-life of its actuators. The design of a suitable pulse transfer function for a plant controller is based on two simple rules and a few standard frequency response curves, which are easily computed once and for all time. Structural resonances are eliminated by digital notch filters, whose pole-zero locations are directly related to the frequency and bandwidth of an oscillatory mode; exactly as with analogue networks (eg. Bridged-Tee). In addition, a computationally simple formula provides an upper-bound on the amplitude of the control error component due to multiplicative rounding effects in the digital computer. The calculation enables the selection of the microprocessor wordlength necessary to meet the control system accuracy specification. A distinct advantage of the proposed comprehensive design technique is its numerical simplicity which eliminates the need for a complex computer-aided design facility.

Time domain synthesis of DDC systems to give dead-beat and minimum control error variance designs are also described in this chapter. An example of a dead-beat system is shown to have a poor step function response, though it ramp response is quite satisfactory. Examination of the corresponding closed-loop frequency response reveals the cause as a resonance which is excited more strongly by the relatively wider frequency spectrum of the step function. The proposed frequency domain technique applied to the same example yields a system with acceptable responses to both step and ramp inputs, and in addition it serves to demonstrate the inter-relationship between the sampling frequency and the achievable closed-loop bandwidth.

3.13 REFERENCES

1. IEE Colloquium: Design of Discrete Controllers, London, December 1977.
2. KNOWLES, J.B. 'A Contribution to Direct Digital Control', Ph D Thesis, University of Mancester 1962.
3. RAGGAZZINI, J.R., FRANKLIN, G.: 'Sampled Data Control Systems', (McGraw Hill,1958).
4. KNOWLES, J.B., EDWARDS, R.: 'Ripple Performance and Choice of Sampling Frequency for a Direct Digital Control System', Proc IEE, 1966, v.133, p.1885.
5. JURY, E.I.: 'Sampled-Data Control Systems', (John Wiley,1958).
6. MATTERA, L.: 'Data Converters Latch onto Microprocessor', Electronics, September 1977 p.81.
7. KNOWLES, J.B., EDWARDS, R.: 'Aspects of Subrate Digital Control Systems', Proc IEE, 1966, v.133, p.1893.
8. AGARWAL, R.C., BURRUS, C.S.: 'New Recursive Digital Filter Structures have a Very Low Sensitivity and Roundoff Error Noise', IEE Trans on Circuits and Systems, 1975, v.22, p.921.
9. WEINSTEIN, C.J.: 'Quantisation Effects in Digital Filters', Lincoln Laboratory MIT Rept.468, November 1969.
10. KNOWLES, J.B., EDWARDS, R.: 'Computatational Error Effects in a Direct Digital Control System', Automatica, 1966 v.4, p.7.
11. KNOWLES, J.B., EWARDS, R.: 'Effect of a Finite Wordlength Computer in a Sampled-Data Feedback System', Proc IEE, 1965, v.112, p.1197.
12. KNOWLES, J.B., EDWARDS, R.: 'Finite Wordlength Effects in Multirate Direct Digital Control Systems', Proc IEE, 1965, v.112, p.2377.
13. WONG, Y.T., OTT, W.E.: (of Burr-Brown Inc). 'Function Circuits - Design and Applications', (McGraw Hill,1976).
14. STEWART, R.M.: 'Statistical Design and Evaluation of Filters for the Restoration of Sampled Data', Proc IRE, 1956, v.44, p.253.
15. KRANC, G.M.: 'Compensation of an Error-Sampled System by a Multi-rate Controller', Trans AIEE, 1957, v.76, p.149.
16. KNOWLES, J.B., EDWARDS, R.: 'Critical Comparison of Multirate and Singlerate DDC system Performance', JACC Computational Methods Session II, 1969.

APPENDIX 3.1   LIST OF SYMBOLS

f Real frequency variable (Hz)

w Real angular frequency (rad/s)

T Sampling Period

f_s Sampling frequency applied to continuous data (Hz) and $f_s \triangleq 1/T$

w_s Angular sampling frequency applied to continuous data (rad/s)

M_p Peak magnification of the closed-loop real frequency response of a DDC system

s Complex frequency variable of Laplace Transformation

$G(s)$ Transfer function of plant

H*(jw)   Pulse real frequency response function of DAC, data hold, and plant

K*(jw)   Overall pulse real frequency response of closed-loop DDC system

ε_R Error signal component due to imperfect attenuation of side-bands generated by sampling continuous input data to a DDC system

D(z)   Pulse transfer function of a digital compensator

K   Scalar gain constant associated with a digital compensator

K_P, K_I Proportional and Integral gain terms in a PI controller

α Attenuation constant in an analogue phase-lag compensator, whose transfer function is R(s)

f_B, w_B Real frequency, angular frequency, at which the phase lag of open loop system is $-130°$

w_o Centre frequency (rad/s) of a digital notch filter with pulse transfer function N(z)

B_{NO} Bandwidth of digital notch filter N(z)

q   Width of data quantization (= $2^{-(wordlength\ -1)}$)

g   Asymptotic gain of plant G(s) -see equation 3.3

R   Rank of plant G(s) -see equation 3.3

Other Variables are defined in the Text

Chapter 4

Self-tuning digital control systems

4.1. INTRODUCTION

The design of direct digital control systems involves a multiplicity of choices concerning sample rates, controller order, control objectives, etc. The designer is thus hard pressed to objectively specify the exact form of a digital feedback controller a-priori. In the same spirit, the synthesis of a digital feedback law often requires more than the hand calculations associated with analogue design. As such, digital controller synthesis can be time-consuming and error-prone when conducted in an off-line context.

With these and related matters in mind the idea of a self-tuning digital control has arisen. The aim of the self-tuning algorithm is to carry out the synthesis, implementation and validation of digital controllers in an on-line, semi-automatic, manner. Such a procedure takes the more routine aspects of controller specification out of the engineers hand and leaves him free to attend to more general qualitative problems. By the same token it allows a range of candidate controllers to be evaluated rapidly, on-line and in a straightforward manner.

In its essential form self-tuning combines the sequence of identification, controller synthesis and implementation as indicated in Fig. 4.1. The sequence is carried out at each sampling interval and proceeds in an iterative manner until the controller coefficients achieve steady values, at which point the identification and synthesis stages are stopped.

As will be clear from this brief explanation, self-tuning is a form of adaptive control. As such it owes its current popularity to the recognition that certain simple identification algorithms could be combined with a regulator synthesis algorithm in an adaptive manner to give a closed-loop system with certain special properties (see Section 5 below). The seminal work [1] used an optimal regulator synthesis procedure, this was followed by a method, again using an optimal cost objective, which allowed combined set-point tracking and disturbance regulation in an elegant manner [2]. The optimal synthesis approach lacks robustness in certain situations, this observation led to the development of classical self-tuners based upon pole-zero assignment concepts [3,4,5,52]. These developments are paralleled by work on the convergence properties of self-tuning algorithms viewed as adaptive stochastic systems [6,7].

The material presented here will concentrate upon one particular form of self-tuner (pole/zero assignment) as follows –

Section 4.2 outlines the nature of the self-tuning algorithm in general terms. Section 4.3 takes the identification phase of the self-tuning sequence and develops the relevant parameter estimation algorithms. The forms of controller synthesis laws which are useful in a self-tuning context are outlined in Section 4.4. The discussion closes with mention of further forms of self-tuning methods and algorithms, together with a brief mention of typical applications.

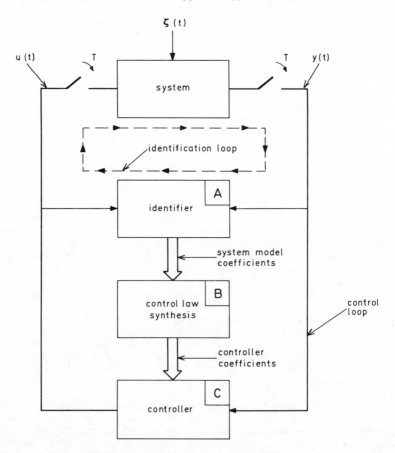

Fig. 4.1 The self-tuning sequence

4.2. SELF-TUNING SEQUENCE

With reference to Fig. 4.1, the system to be controlled is assumed to be a linear finite order system which can be modelled by a z transform description of the form

$$y(t) = \frac{z^{-k} B(z^{-1})}{1+A(z^{-1})} u(t) + \frac{C(z^{-1})}{1+A(z^{-1})} \xi(t) \qquad (4.1)$$

where $u(t)$, $y(t)$ $(t = 0,\pm1,\pm2,...)$ are the time <u>sequences</u> associated with the controller output and the sampled system output, and $\xi(t)$ is a white noise sequence. The transfer function which is characterised by the polynomials

$$\left. \begin{aligned} B(z^{-1}) &= b_1 z^{-1} + \ldots + b_{n_b} z^{-n_b} \\ A(z^{-k}) &= a_1 z^{-1} + \ldots + a_{n_a} z^{-n_a} \end{aligned} \right\} \qquad (4.2)$$

represents the input/output dynamics of the system to be controlled and is related to the underlying continuous time system by the normal z-transform relations, where the zero-order hold associated with the controller analogue output port is considered part of the underlying continuous system. Any time delays associated with the input/output dynamics, which are integer multiples of the sample interval T are modelled by the term z^{-k}, while delays which are fractions of T contribute an additional discrete time zero (see table 1 for an appropriate set of z transforms and [5] for a discussion).

The transfer function which is characterised by $A(z^{-1})$ and the polynomial

$$C(z^{-1}) = 1 + c_1 z^{-1} + c_2 z^{-2} + \ldots + c_{n_c} z^{-n_c} \qquad (4.3)$$

represents the aggregate influence of random disturbances on the system output.

Under the conditions which prevail in self-tuning the complex variable z^{-1} can be interpreted [8] as the unit delay operator, so that eqn. 4.1 can be written as a difference equation thus

$$y(t) = -\sum_{i=1}^{n_a} a_i\, y(t-i) + \sum_{j=1}^{n_b} b_j\, u(t-j-k) + \varepsilon(t) \qquad (4.4)$$

where for convenience we have written the disturbance as

$$\varepsilon(t) = \xi(t) + c_1\, \xi(t-1) + \ldots + c_{n_c} \xi(t-n_c) \qquad (4.5)$$

The aim of the self-tuning identifier (section 4.3) is to determine the coefficients $\{a_i, b_j\}$ of the polynomials $A(z^{-1})$, $B(z^{-1})$ which describe the unknown system. At each sample interval these coefficients are re-estimated (by updating previous estimates) and passed on to the controller synthesis block. The control synthesizer obtains from the system model coefficients a set of controller coefficients $\{f_i, g_i\}$ which are then passed to the controller which, in the simplest regulatory formulation, is defined by

$$u(t) = -\frac{G(z^{-1})}{1+F(z^{-1})}\, y(t) \qquad (4.6)$$

where

$$\left. \begin{aligned} F(z^{-1}) &= f_1 z^{-1} + \ldots + f_{n_f} z^{-n_f} \\ G(z^{-1}) &= g_0 + g_1 z^{-1} + \ldots + g_{n_g} z^{-n_g} \end{aligned} \right\} \qquad (4.7)$$

Related controller structures for servo-tracking will be discussed in

Section 4.4 below, for the moment it is important to notice that –

(a) The controller does indeed <u>synthesize</u> the controller coefficients by solving a set of linear equations which has a unique solution. There is no element of design (implying a multiplicity of choices) in this stage of the self-tuner.

(b) The difference equation model of the system can sometimes be reticulated in such a manner that the synthesis rule becomes scrambled up with the identifier. Nevertheless, a synthesis law is always buried somehwere in the computations (see [17] and [52] for a discussion of this point).

The sequence of identification, synthesis and implementation is started using some approximate model of the system, and at each sample interval the corrective nature of the identifier updates the controller coefficients in a manner which moves progressively toward the correct controller configuration. Again note that the procedure is inherently auto–synthetic, the design aspect of self-tuning takes place at a higher level and involves engineering judgement as to the following matters –

i)   <u>Sample interval, T.</u> No algorithmic procedure exists for determining a 'best' sampling interval (see [9] for a discussion).

ii)   <u>System model order, n_a, n_b, k.</u> Again no acceptable objective procedure exists here, although guidance is available through a variety of techniques (see [18,19] for example).

iii) <u>Control objective.</u> Here the basic choice is between the various synthesis rules to be incorporated in block B of Fig. 4.1 (see Section 4.4 below).

In addition the engineer must also make the more fundamental decision as to whether his problem is one of stochastic regulation, servo–input tracking, or a mix of the two. For further discussion of these issues see [10] and Section 4.4 of this article.

4.3. SYSTEM IDENTIFICATION

As indicated in the previous section, the identifier involves an updating procedure whereby the data which become available at the beginning of each sample interval are used to improve the previous estimate of the system model. Such estimators are termed <u>recursive,</u> and the literature contains many techniques which fit this description. In self-tuning, the most commonly used recursive estimator is based upon least squares estimation [11,12]. The recursive version of least squares can take a variety of forms [13], however the one most usually considered is the original method due to Plackett [14], since it is easy to programme and works well in a variety of situations (the numerical properties of this method make it unsuitable for fixed–point or short–word–floating–point computation. See [12,19] and general references to square root and upper diagonalisation methods for numerically superior methods).

Consider the difference equation representation of eqn. 4.4, this can be written in vector matrix format as –

$$y(t) = z^T(t)\underline{\theta} + \varepsilon(t) \qquad (4.4 \text{ bis})$$

where $\quad z^T(t) = (-y(t-1), \ldots, -y(t-n_a), u(t-1-k), \ldots u(t-n_b-k))$

$\quad\quad \underline{\theta}^T = (a_1, \ldots, a_{n_a}, b_1, \ldots, b_{n_b})$

Then the recursive least square estimate of the vector of system coefficients $\underline{\theta}$ at time interval t is given by [14]

$$\tilde{\underline{\theta}}_t = \tilde{\underline{\theta}}_{t-1} - \underline{P}_t \underline{z}(t) \, (\underline{z}^T(t) \, \tilde{\underline{\theta}}_{t-1} - y(t)) \quad\quad (4.8)$$

where $\quad \underline{P}_t = \underline{P}_{t-1} - \underline{P}_{t-1} \, \underline{z}(t) \, \underline{z}^T(t) \, \underline{P}_{t-1} \, (1 + \underline{z}^T(t) \, \underline{P}_{t-1} \, \underline{z}(t))^{-1} \quad\quad (4.9)$

Notice that eqn. (4.8) is a recursion for the estimated vector of coefficients $\tilde{\underline{\theta}}_t$ at time t, based upon those at time step t-1. Taken with eqn. (4.9) the recursion operates as follows at time step t -

(a) Calculate $\underline{P}_t$ based upon $\underline{P}_{t-1}$ and the new data u(t) and y(t).

(b) Calculate $\tilde{\underline{\theta}}_t$ based upon $\underline{P}_t$, $\tilde{\underline{\theta}}_{t-1}$ and the new data u(t) and y(t).

The following points are relevant.

i)   The algorithm is related to a Kalman Filter for a system with constant states [14].

ii) The algorithm is a member of the class of stochastic gradient methods which find general expression in stochastic approximation methods [20].

iii) The algorithm does not involve a matrix inversion (as does off-line least squares) yet it yields a vector of estimated coefficients.

iv) At time step t=1, the initial matrix $\underline{P}_0$ and vector $\underline{\theta}_0$ must be specified. One can use guesses of coefficients in $\tilde{\underline{\theta}}_0$ and set $\underline{P}_0$ to a diagonal matrix with large coefficients down the diagonal. (See [10,15,22,23,49] for discussions of this point in relation to self-tuning).

In self-tuning it is sometimes required to track time-varying coefficients in the system. The recursive least squares estimator is given the required tracking feature by means of a 'forgetting factor' [14]. The idea is to use the fact that the matrix $\underline{P}_t$ reflects our certainty concerning the estimates. It is, in fact, related to the statistical covariability of the estimated coefficients. The tracking facility is obtained by shortening the memory of the estimator, and this is done by artificially boosting $\underline{P}_t$ at each time step. Thus eqn. (4.9) is rewritten as -

$$\underline{P}_t = \bar{\underline{P}}_{t-1} - \bar{\underline{P}}_{t-1} \, \underline{z}(t) \underline{z}^T(t) \, \bar{\underline{P}}_{t-1} \, (1 + \underline{z}^T(t) \bar{\underline{P}}_{t-1} \underline{z}(t))^{-1} \quad\quad (4.10)$$

where $\quad \bar{\underline{P}}_{t-1} = \dfrac{\underline{P}_{t-1}}{\rho} \quad\quad\quad\quad (4.11)$

The 'forgetting factor' ρ is set between 0 and 1, and controls the speed with which the estimates $\tilde{\underline{\theta}}_t$ can adapt to changes in the system. As ρ is decreased the speed of adaptation increases, but this adaptive capability is gained at the expense of increased statistical uncertainty and values of ρ below 0.95 are rarely used.

An alternative means of incorporating tracking capability into the recursive estimator is to use a 'random walk' method whereby eqn. (4.10) is used to construct $\underline{P}_t$ but $\overline{\underline{P}}_{t-1}$ is defined by

$$\overline{\underline{P}}_{t-1} = \underline{P}_{t-1} + \underline{R} \qquad (4.12)$$

where $\underline{R}$ is a positive definite matrix whose entries can be varied to allow the identifier to track changes in the system. Further discussion on the use of such techniques in self-tuning is given in [10,21,43,27].

Various other identification algorithms have been proposed for self-tuning, but none is as effective as least squares. It was hoped that computationally 'fast' algorithms for performing the least squares recursion would prove advantageous. Unfortunately these algorithms have not proven as useful as was expected (see [13]).

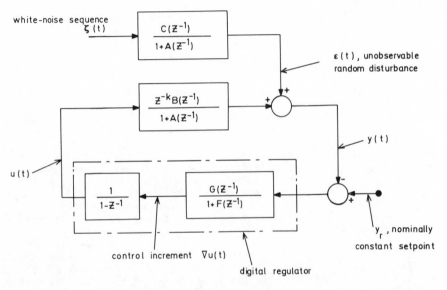

Fig. 4.2 Stochastic regulation

4.4. CONTROLLER SYNTHESIS

Before discussing specific algorithms it is as well to distinguish the form of feedback control action which is required. First, if the problem is one of regulating an output $y(t)$ in the presence of stochastic disturbances such that it corresponds as closely as possible to a nominally constant value y_r, then the configuration indicated in Fig. 4.2 is appropriate. Notice that a digital integrator is incorporated to ensure steady state correspondence (in the sense of expectation) between $y(t)$ and y_r. The control objective is the reduction of the influence of the extraneous disturbance $\varepsilon(t)$ on the output $y(t)$. The controller is therefore correctly referred to as a regulator, although in practice such a configuration deals effectively with slow changes in set-point as well.

If servo-tracking of a (relatively) quickly changing reference signal $y_r(t)$ is required, and the stochastic disturbance $\varepsilon(t)$ is small then the design focus switches to the <u>servo-tracking</u> of $y_r(t)$ in some appropriate way, rather than rejecting the stochastic disturbance. The difference is one of emphasis, since as mentioned above, the correctly designed regulator with integral action will in the steady state track changes in reference signal. Nevertheless, the feedback law is now correctly referred to as a servo-controller and could look like the system in Fig. 4.3, where the only difference is in the values of the coefficient of $F(z^{-1})$ and $G(z^{-1})$ and the addition of the precompensator $S(z^{-1})$ which shapes the transient changes in the reference command before it hits the system. In practice, any well designed DDC system would incorporate transient precompensation, if only in the form of 'rate-limiting' of step demands (see [10] and [15] for a discussion of this point).

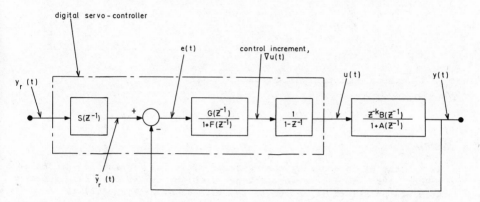

Fig. 4.3 Servo control

If the control situation demands a combination of servo-tracking of a rapidly or frequently changing reference signal <u>and</u> rejection of significant stochastic disturbances then the configuration of Fig.4.4 is useful [2,52]. Again the difference between this and Figs. 4.2 and 4.3 is minimal, indeed this last arrangement could be used as a prototype configuration. The distinctive feature of the layout is the composite control objective of set-point tracking and disturbance rejection, which leads to different control law coefficients than would arise if the aim were only regulation or only servo-tracking.

In addition to differentiating between the form of control action as above, it is also necessary to establish a design philosophy. In self-tuning terms this amounts to the alternative of an optimal control objective or a classical control objective [16]. The latter, which is discussed in detail here, is associated with pole/zero assignment and has qualities of closed-loop integrity which are not necessarily present in the optimal control self-tuner [5]. Indeed most research groups have now adopted pole-assignment as the most robust algorithm available.

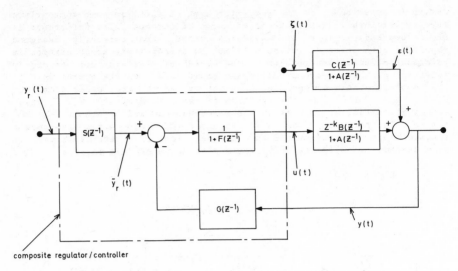

composite regulator / controller

Fig. 4.4 Composite servo/regulation configuration

Example

As a means of introducing the general form of synthesis procedure consider the simple servo-tracking problem in which a second order system has been identified (using the methods of Section 4.2) as

$$W(z^{-1}) = \frac{z^{-k}B(z^{-1})}{1+A(z^{-1})} = \frac{z^{-1}(b_1+z^{-1}b_2)}{1+a_1z^{-1}+a_2z^{-2}} \tag{4.13}$$

Assume incremental control (as per Fig. 4.3) such that the digital integrator is lumped with the system, thus

$$y(t) = \frac{W(z^{-1})}{1-z^{-1}} \Delta u(t) = W'(z^{-1}) \Delta u(t) \tag{4.14}$$

where

$$W'(z^{-1}) = \frac{z^{-1}(b_1+z^{-1}b_2)}{1+\alpha_1z^{-1}+\alpha_2z^{-2}+\alpha_3z^{-3}} \tag{4.15}$$

with

$$\left.\begin{array}{l} \alpha_1 = a_1-1 \\ \alpha_2 = a_2-a_1 \\ \alpha_3 = -a_2 \end{array}\right\} \tag{4.16}$$

Propose the control law (Fig. 4.3) given by

$$\Delta u(t) = \left\{ \frac{g_0+z^{-1}g_1+z^{-2}g_2}{1+f_1z^{-1}} \right\} e(t) \tag{4.17}$$

and

$$e(t) = \tilde{y}_r(t) - y(t) = S(z^{-1})y_r(t) - y(t) \tag{4.18}$$

Combining the control law (eqn. 4.17) and the system equation (eqn. 4.14) gives the closed-loop equation

$$y(t) = \frac{z^{-1}(g_0+z^{-1}g_1+z^{-2}g_2)(b_1+z^{-1}b_2)S(z^{-1})\ y_r(t)}{D(z^{-1})} \qquad (4.19)$$

where the denominator $D(z^{-1})$ is given by -

$$D(z^{-1})=(1+a_1z^{-1}+a_2z^{-2}+a_3z^{-3})(1+f_1z^{-1})+z^{-1}(b_1+z^{-1}b_2)(g_0+z^{-1}g_1+z^{-2}g_2)$$

$$(4.20)$$

Now, $D(z^{-1})$ is the closed-loop characteristic polynomial of the servo-system, thus if we wish to <u>specify</u> desired closed-loop pole positions corresponding to the roots of

$$T(z^{-1}) = 1+t_1z^{-1}+t_2z^{-2}+t_3z^{-3} \qquad (4.21)$$

Then we must synthesize the appropriate controller coefficients by equating $D(z^{-1})$ and $T(z^{-1})$ and solve the set of equations obtained by equating like powers of z^{-1}. These equations are

$$\begin{bmatrix} 1 & b_1 & 0 & 0 \\ a_1 & b_2 & b_1 & 0 \\ a_2 & 0 & b_2 & b_1 \\ a_3 & 0 & 0 & b_2 \end{bmatrix} \begin{bmatrix} f_1 \\ g_0 \\ g_1 \\ g_2 \end{bmatrix} = \begin{bmatrix} t_1 - a_1 \\ t_2 - a_2 \\ t_3 - a_3 \\ 0 \end{bmatrix} \qquad (4.22)$$

which can be easily solved for the controller coefficients, (f_1,g_0,g_1,g_2). Now if $S(z^{-1})$ is chosen to have $(g_0+g_1z^{-1}+g_2z^{-2})$ as a denominator (or the modified implementation of Fig.4.5 is used) then the closed-loop equation is

$$y(t) = \frac{z^{-1}(b_1+z^{-1}b_2)}{T(z^{-1})}\ y_r(t) \qquad (4.23)$$

which has the desired pole set, but still has the zeros of the open loop system. Now zeros cannot be shifted by feedback, so if it is required to remove these zeros directly then they must be cancelled. The process of cancelling system zeros is implicit in optimal control type self-tuning [1,2] and is inherently dangerous in digital control because the $B(z^{-1})$ polynomial is <u>very likely</u> to be non-minimum phase (i.e. roots outside the unit disc) such that unstable control results from an attempt at cancellation. This crucial practical observation is the raison dêtre for the classically derived self-tuners. The danger of zero-cancellation can be seen in terms of the simple example by selecting the controller pole f_1 to be

$$f_1 = \frac{b_2}{b_1} \qquad (4.24)$$

This causes the zero to be cancelled by the imposed common factor in the denominator. The controller zeros are then found from,

$$(1+a_1 z^{-1}+a_2 z^{-2}+a_3 z^{-3})+z^{-1}b_1(-g_0+g_1 z^{-1}+g_2 z^{-2}) = T(z^{-1}) \qquad (4.25)$$

to be

$$g_0 = b_1^{-1}(t_1-a_1)$$

$$g_1 = b_1^{-1}(t_2-a_2) \qquad (4.26)$$

$$g_2 = b_1^{-1}(t_3-a_3)$$

Notice that as b_1 decreases relative to b_2 the controller pole moves outside the unit disc and the closed-loop system becomes unstable.

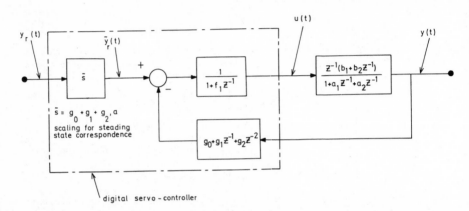

Fig. 4.5 Final closed-loop configuration for illustrative example

Algorithms Operating on the System

The previous example indicates the principle of the synthesis method employed in self-tuning, albeit in terms of a simple servo-tracking system. Consider now the general algorithm for servo-tracking or stochastic regulation, obtained by rewriting equation (4.1) as

$$y(t) = \frac{z^{-k}B(z^{-1})}{1+A(z^{-1})}\Delta u(t) + \frac{C(z^{-1})}{1+A(z^{-1})}\xi(t) \qquad (4.27)$$

Where the incremental control action $\Delta u(t)$ has been incorporated into the basic system description and the polynomial $A(z^{-1})$ extended accordingly (cf. the illustrative example).

Propose the feedback law

$$\Delta u(t) = \frac{G(z^{-1})}{1+F(z^{-1})}(y_r-y(t)) \qquad (4.28)$$

Where for the synthesis procedure to have a unique solution (cf. eqn. (4.22)) the orders of n_f and n_g must be set to

$$n_g = n_a-1, \; n_f = n_b+k-1 \qquad (4.29)$$

By combining the eqns. (4.27) and (4.28) the closed-loop equation is found to be

$$\{(1+A)(1+F) + z^{-k}BG\}y(t) = z^{-k}BG\ y_r(t) + C(1+F)\xi(t) \qquad (4.30)$$

where the explicit dependence upon z has been dropped from the notation. If the stochastic disturbance is negligible then the servo-tracking control law is found by solving the identity

$$\{(1+A)(1+F) + z^{-k}BG\} = T_1 \qquad (4.31)$$

for the unknown controller coefficients. The set of equations obtained from (4.31) is the general version of equation set (4.22), and $T_1(z^{-1})$ is a polynomial which defines the desired pole set.

If the aim is to obtain good closed-loop regulation properties against the disturbance $\varepsilon(t)$, then the regulator coefficients are obtained by solving the identity

$$\{(1+A)(1+F) + z^{-k}BG\} = CT_2 \qquad (4.32a)$$

Note that the time delay through the system can be ignored by setting k to zero and extending the order of $B(z^{-1})$ by k_{max}, where k_{max} is the maximum anticipated time delay in the system [5].

The modified regulator coefficients are now calculated as the solution to
$$\{(1+A)(1+F) + BG\} = CT_2 \qquad (4.32b)$$

Where again $T_2(z^{-1})$ is a polynomial which defines the closed-loop pole set and the noise polynomial is included on the right hand side of eqn. (4.32) so that it can be cancelled in the closed-loop equation, (recall that $C(z^{-1})$ is a realisation of a stochastic process and can always be made minimum phase so that cancellation is permitted. See [24] for a discussion of this point) to give regulation error

$$y(t) = \frac{(1+F)\xi(t)}{T_2} \qquad (4.33)$$

Notice that because of the incremental control action steady state correspondence between $y_r(t)$ and $y(t)$ is assured. However, the shape of the transient response can be poor since it is governed by the transfer function −

$$\frac{z^{-k}BG}{T_2C}\ y_r(t)$$

Nevertheless, the robust regulatory performance of controller coefficients calculated from eqn. (4.32) commends this particular approach. Moreover, practical experience suggests that rate limited reference inputs provide an acceptable servo-input response even when the design is based upon disturbance regulation (see [3,5] for a further discussion of these points).

If optimal regulation is required the synthesis rule must be altered to cancel $B(z^{-1})$ and take account of the time delay z^{-k} through the system. The latter is done by rewriting the system (eqn. 4.27) as

$$y(t) = \frac{z^{-k}B'(z^{-1})}{1+z^{-k}A'(z^{-1})} \Delta u(t) + \frac{C(z^{-1})P_A(z^{-1})}{1+z^{-k}A'(z^{-1})} \xi(t) \tag{4.34}$$

where $P_A(z^{-1}) = 1+p_1 z^{-1} +...+p_k z^{-(k-1)}$ is a polynomial with coefficients which cancel out the leading $k-1$ coefficients of $A(z^{-1})$ and

$$1+z^{-k}A'(z^{-1}) = (1+A(z^{-1}))P_A(z^{-1}) = 1+z^{-k}(a_1' z^{-1} + ... + a_{n_a}' z^{-n_a})$$

$$B'(z^{-1}) = b_1' z^{-1} + ... + b_{n_b+k}' z^{-n_b-k} = B(z^{-1})P_A(z^{-1})$$

Substituting for the feedback law (eqn. 4.28) gives the closed-loop equation

$$[(1+z^{-k}A')(1+F) + z^{-k}B'G]y(t) = z^{-k}GB'y_r(t) + (1+F)(M+z^{-k}C')\xi(t) \tag{4.35}$$

where $P_A(z^{-1})C(z^{-1}) = M(z^{-1}) + z^{-k}C'(z^{-1})$

and $M(z^{-1})$ is the polynomial consisting of the first k terms of $P_A(z^{-1})C(z^{-1})$.

The minimum variance (optimal) regulator [24,3] coefficients are then given by the solution to

$$M(z^{-1})(A'(z^{-1})(1+F(z^{-1})) + B'(z^{-1})G(z^{-1})) = (1+F(z^{-1}))C'(z^{-1}) \tag{4.36}$$

which leaves the regulation error a kth order moving average [24] as required

$$y(t) = M(z^{-1}) \xi(t) \tag{4.37}$$

The important thing about eqn. 4.36 is that if the polynomial $C(z^{-1})$ is unity (white disturbance) then $C'(z^{-1}) = 0$ and the solution to equation 4.37 becomes

$$\left. \begin{aligned} G(z^{-1}) &= z\, A'(z^{-1}) \\ (1+F(z^{-1})) &= \frac{z\, B'(z^{-1})}{b_1'} \end{aligned} \right\} \tag{4.38}$$

It is this simple fact that the rearranged system parameters are the optimal regulator coefficients (for $C(z^{-1}) = 1$) that is used in implicit self-tuning. Also important is the 'k step predictor' format for the system description (eqn. 4.34) whereby the current output is expressed in terms of ($y_{t-k-1}, y_{t-k-2}, ...$ and $u_{t-k-1}, u_{t-k-2}...$). For further discussion of this point and relations between pole/zero allocation and optimal regulation see [3,5] and books on linear quadratic regulation, such as that by Anderson and Moore. For a discussion of servo-tracking synthesis rules see [17], and note the complimentary nature of the various algorithms for servo-tracking [17] and regulation [3]. This point is carried further in Section 4.8.

Algorithms Operating on a Pseudo-system

In the foregoing and elsewhere [16] it has been emphasised that optimal control based synthesis algorithms involve cancellation of the system numerator $B(z^{-1})$. Furthermore, because $B(z^{-1})$ is likely to be non-minimum phase, this procedure can lead to very poor or even unstable systems. An alternative procedure to the classically derived [25] synthesis methods mentioned above is to change the open-loop system zeros by feedforward (see [47] for a discussion of this point) and then cancel the minimum phase zeros of the pseudo-system which is so obtained. This is one interpretation of the algorithms developed by Clarke et. al. at Oxford [2,26], whereby the system has feedforward applied to it and has its output filtered (Fig. 4.6) such that the zeros of the pseudo-system so obtained can (with correctly selected feedforward) be safe for cancellation. This approach has the attractive feature that servo-tracking <u>and</u> regulation can be combined in a composite linear quadratic cost function. To be specific, the output of the pseudo-system (Fig. 4.6) is given by

$$\emptyset(t) = P(z^{-1})y(t) + Q(z^{-1})u(t-k) - R(z^{-1})y_r(t-k) \qquad (4.39)$$

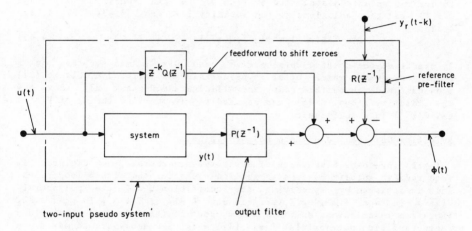

Fig. 4.6 Block diagram of pseudo-system for self tuning

The composite control aim is to minimise the mean square value of $\emptyset(t)$, where the inclusion of the servo-input and its shaping filter $R(z^{-1})$ give the reference signal tracking property and the mean square weighting on $y(t)$ offers a reduction in the output variance due to the disturbance. The synthesis method works by putting eqn. 4.39 and the system (eqn. 4.1) into a generalised form of the 'k-step predictor' format mentioned earlier thus

$$\emptyset(t) = G(z^{-1})y(t-k)+H(z^{-1})y_r(t-k) + (1+F(z^{-1}))u(t-k) + \varepsilon(t) \qquad (4.40)$$

Where $\varepsilon(t)$ is arranged to be a moving average noise sequence of order k-1 as required to minimize the variability of $\emptyset(t)$. The pseudo output $\emptyset(t)$ is then made equal to the noise sequence by applying the control sequence (Fig. 4.4)

$$u(t) = -F(z^{-1})u(t) - G(z^{-1})y(t) - H(z^{-1})y_r(t) \qquad (4.41)$$

where
$$1+F(z^{-1}) = B(z^{-1})E(z^{-1}) + Q(z^{-1})C(z^{-1}) \qquad (4.42a)$$

$$H(z^{-1}) = -R(z^{-1})C(z^{-1}) \qquad (4.42b)$$

and the polynomials $G(z^{-1})$ and $E(z^{-1})$ are defined by the identity

$$C(z^{-1})P(z^{-1}) = E(z^{-1})A(z^{-1}) + q^{-k}G(z^{-1}) \qquad (4.43)$$

This expression should be compared with the standard form synthesis rules illustrated earlier (eqns. 4.31, 4.32a, 4.32b).

Notice that if $Q(z^{-1}) = 0$, then from eqn. 4.42a, $B(z^{-1})$ appears in the denominator of the controller and will cause stability problems for the reasons given earlier. In fact by manipulation of eqns. 4.43, 4.42, 4.41 and 4.1 the closed—loop equation can be shown to be

$$y(t) = \frac{z^{-k}B(z^{-1})R(z^{-1})}{A(z^{-1})Q(z^{-1})+B(z^{-1})P(z^{-1})} y_r(t) + \frac{(1+F(z^{-1}))\xi(t)}{A(z^{-1})Q(z^{-1})+B(z^{-1})P(z^{-1})} \qquad (4.44)$$

Thus in order to move the discrete time poles of the closed—loop system to desired positions specified by $T_3(z^{-1})$ we can specify $P(z^{-1})$ and $Q(z^{-1})$ as the solutions to the identity (cf. eqns. 4.43, 4.31, 4.32a, 4.32b).

$$P(z^{-1})B(z^{-1}) + A(z^{-1})Q(z^{-1}) = T_3(z^{-1}) \qquad (4.45)$$

In practice, the coefficients of $P(z^{-1})$ and $Q(z^{-1})$ can be set in a number of ways which offer considerable design freedom. Clarke [27] provides guidance on this point and Allidina and Hughes [28] show how the identity (eqn. 4.45) can be used to give on—line tuning of $P(z^{-1})$ and $Q(z^{-1})$ in a self-tuner.

Selecting Desired Closed—Loop Pole Positions

In the foregoing discussion of synthesis procedures the designer is required to specify desired target locations for the closed—loop poles. This is achieved by specifying the coefficients of the polynomials $T(z^{-1})$ in eqn. 4.21, or $T_1(z^{-1})$ in eqn. 4.31, or $T_2(z^{-1})$ in eqn. 4.32. There are restrictions on the maximum order of these desired closed—loop characteristic polynomials (see [3]) which are necessary to make sure that the corresponding identity has a solution. In practice, however, the desired closed—loop pole-set is best specified in terms of either a first or second order polynomial in z. This in turn can be reflected back through the inverse z-transformation to relate to the desired bandwidth and transient response of the <u>closed-loop continuous time response</u>. It is therefore important to realise that the discrete time desired pole set must be specified in conjunction with the sample interval T. It is a power of the self-tuning approach that the desired pole set and sample interval T can be set-up experimentally and various schemes quickly tried [10,31]. However, for general preliminary guidance the following points have been found useful.

i) Decide upon the achievable closed—loop bandwidth of the continuous system f_c, bearing in mind the various engineering constraints upon speed of response.

ii) Pick a sample rate three or four times greater than f_c. That is –

$$T \approx 1/4f_c$$

iii) Specify a first or second order desired closed—loop pole set and determine the coefficients of the discrete time characteristic polynomial by using a z transformed version of the desired continuous time response.

For a first order response

$$T(z^{-1}) = 1 + z^{-1}t_1$$

where $t_1 = -\exp(-T/\tau)$ and τ is the desired closed-loop time constant, and typically $\tau \approx 3T$.

For a second order response

$$T(z^{-1}) = 1 + z^{-1}t_1 + z^{-2}t_2$$

where

$$t_1 = -2e^{-\xi w_n T} \cos (w_n(\sqrt{1-\xi^2}T))$$

$$t_2 = e^{-2\xi w_n T},$$

and typically $w_n\sqrt{1-\xi^2} \, T \approx 1/3$.

The best approach to setting sample interval and closed-loop pole set is an interactive self-tuning manner in which various combinations are tried, each successive combination attempts to speed-up the system performance further, until a natural limit is found.

4.5. BASIC SELF-TUNING PROPERTY [1,2,3,4]

The popularity of self-tuning is due to a theoretical result which makes on-line synthesis simpler than the off-line equivalent. To be more specific the self-tuning property tells us that even if the stochastic disturbance $\varepsilon(t)$ is heavily coloured $(C(z^{-1})\neq1)$ we can assume it to be white $(C(z^{-1})=1)$ and still get the correct closed-loop control law, provided the control law coefficients are set by the self-tuning configuration of Fig.4.1. In computational terms and for the simple regulator $(y_r = 0)$ with a pole assignment objective this means at each sample interval we:-

(a) apply recursive least squares (section 4.3) to update a model of the system based upon the (incorrect) assumption that the stochastic disturbance is white. Suppose the model at time step t is

$$y(t) = -\tilde{A}(z^{-1})y(t) + \tilde{B}(z^{-1})u(t) + \xi'(t) \qquad (4.46)$$

(b) Use the estimated coefficients $\tilde{A}(z^{-1})$ and $\tilde{B}(z^{-1})$ to synthesise regulator coefficients via the identity (cf. eqn. 4.32b)

$$(1+\tilde{A}(z^{-1}))(1+F(z^{-1})) + \tilde{B}(z^{-1})G(z^{-1}) = T_2(z^{-1}) \qquad (4.47)$$

where $T_2(z^{-1})$ defines the desired target pole set.

(c) Apply a control input u(t) determined from

$$u(t) = - \frac{G(z^{-1})}{1+F(z^{-1})} \, y(t) \qquad (4.48)$$

The 'self-tuning property' states that if the above cycle of events converges then the correct control configuration (which would have been calculated off-line with eqn. 4.32b) is a possible convergence point. This self-tuning property can be proven for optimal and classical synthesis laws. A nice feature of the algorithms is that the fitting residual $\xi'(t)$ is related to the actual noise $\xi(t)$ which drives the stochastic disturbance. In the pole-assignment cycle [5] outlined above

$\xi'(t) \to \xi(t)$, indicating the close link between self-tuning and approximate recursive methods for estimating coefficients of $C(z^{-1})$. (See [4], [29] and the discussion of extended least squares in [30] and Appendix 1) most of which rely upon obtaining an estimate of the driving white noise.

The self-tuning cycle may not converge (although in practice so much software for protection of the system and controller is used [31] that convergence of some kind can usually be obtained), so that much theoretical work is devoted to convergence proofs (see [6,7] and more generally, [32]). However, in this discussion it is assumed that convergence is not a problem.

The key point is that the desired closed-loop configuration is obtained without a direct knowledge of $C(z^{-1})$ which would require a maximum likelihood estimator in the off-line case [45]. In the pole-assignment case the factor $C(z^{-1})$ is dropped from the right hand side of eqn. (4.47) (cf. eqn. 4.32b). For minimum variance self-tuning [1] a 'k-step predictor' format estimator is used (cf. eqn. 4.34)

$$y(t) = \tilde{A}(z^{-1})y(t-k) + \tilde{B}(z^{-1})u(t-k) + \xi'(t), \qquad (4.49)$$

in the identification stage. No explicit synthesis occurs, instead the control law is set as –

$$u(t) = - \frac{\tilde{A}(z^{-1})}{\tilde{B}(z^{-1})} \, y(t) \qquad (4.50)$$

Compare with eqn. 4.38, and observe that the self-tuning process is here one of setting $y(t) = \xi'(t)$ at each time step.

The pseudo-system approach of Clarke uses the same principle but with the extended model (compare eqn. 4.40):–

$$\emptyset(t) = \tilde{G}(z^{-1})y(t-k) + \tilde{H}(z^{-1})y_r(t-k) + \tilde{F}(z^{-1})u(t-k) + \xi'(t) \qquad (4.51)$$

and applying control at each step

$$u(t) = \frac{\tilde{G}(z^{-1})}{\tilde{F}(z^{-1})} \, y(t) - \frac{\tilde{H}(z^{-1})}{\tilde{F}(z^{-1})} \, y_r(t), \qquad (4.52)$$

such that $\emptyset(t)$ is forced equal to $\xi'(t)$. Again notice how a parallel with approximate maximum likelihood is formed since $\tilde{H}(z^{-1})$ will contain $C(z^{-1})$ as a factor (equation 4.42b). (See also [4] for a discussion of this point).

The servo-controller algorithm (equation 4.31 and the illustrative example of section 4.4) does not require a knowledge of $C(z^{-1})$ since the extraneous disturbance is negligible. This has advantages, but is also problematic, since there is no noise to 'feed' the estimator stage of the self-tuner. As a result special care is needed in the identification stage [17] to prevent things going wrong.

4.6. TIME-VARYING PARAMETERS AND TIME-DELAYS

Self-tuning allows rapid on-line design and testing of DDC loops and in the current context this must be seen as its key advantage over the off-line cycle of design-implementation-validation. However, with some modification to the estimation phase, self-tuning can be made parameter adaptive. The modification involves the use of a limited memory identifier which allows the estimated system coefficients to change and hence reflect time-varying dynamics. The memory of a self-tuner is most easily altered by a forgetting factor (section 4.3) and this can be automated as per [21]. Alternatively, modification of identifier memory can be seen as an interactive design parameter as shown in [31] and [10].

Time-varying parameters are often reflections of non-linearities in a system which cause the locally linearized transfer function to change with the demanded operating level [15], as such all self-tuners deal with non-linearities of this kind as a parameter variation. The question of fluctuations in system time-delay is more difficult, since this is partially a variation in model order (integer k) and parameters (the coefficients of $B(z^{-1})$ which are a function of the fractional system time delay).

The pole-assignment approach can successfully deal with this situation, by neglecting the time delay and estimating the leading k coefficients which in theory are zero [5]. Optimal control self-tuners use a 'k step predictive' model (see section 4.4 eqn. 4.34 and section 4.5, eqns. 4.49, 4.51), therefore an a priori knowledge of k is built into the formulation. Thus, by virtue of the requirement to know k and the non-minimum phase behaviour induced by time-delays in the underlying system, optimal self-tuners have sensitivity problems which need to be researched more fully.

4.7. MULTIVARIABLE SELF-TUNERS

It is generally true to say that multivariable versions of self-tuning algorithms require significant further work. However, some basic algorithms are available. In particular, see [33,34] for optimal (minimum variance) multivariable regulators. For the multivariable version of the pole-assignment self-tuner see [35,49] together with statements regarding its robustness which parallel the previous discussion of single-variable algorithms.

The essential problem in multivariable self-tuning is the non-commutativity of the various polynomial matrices. In [33] and [35] this is solved using 'pseudo-commutivity' relations of the form

$$\underline{G}(z^{-1}) \ \underline{F}^{-1}(z^{-1}) = \underline{\tilde{F}}^{-1}(z^{-1})\underline{\tilde{G}}(z^{-1}) \qquad (4.53)$$

where all terms in eqn. 4.53 are polynomial matrices. For suitably dimensioned matrices eqn. 4.53 allows the right hand divisor pair $G(z^{-1})\underline{F}^{-1}(z^{-1})$ to be replaced by a different left hand divisor pair $\underline{\tilde{F}}^{-1}(z^{-1})\underline{\tilde{G}}(z^{-1})$. Moreover, the eqn. 4.53 which allows this pseudo-commutivity also provides an identity (compare with the identities used in the single variable case)

$$\widetilde{\underline{F}}(z^{-1})\underline{G}(z^{-1}) - \widetilde{\underline{G}}(z^{-1})\underline{F}(z^{-1}) = 0, \qquad (4.53 \text{ bis})$$

which can be used on-line to get the pseudo-commuted matrix polynomials, $\widetilde{\underline{G}}(z^{-1})\widetilde{\underline{F}}(z^{-1})$.

Further problems exist concerning the structure of the multivariable system model. The approach in [35] is to assume that the system is generic [36] and continue accordingly. As it turns out the method works well in practice and has good properties of robustness. In particular, it is in the area of time-delays and non-minimum phase behaviour where pole-assignment scores heavily. Minimum variance (optimal) derived multivariable self-tuners require that (except in certain cases [33]) all the system input channels have the same time delay. Clearly this restricts the usefulness of the optimal control approach. Likewise difficulties with non-minimum phase properties become much more severe in the multivariable case.

4.8. EXTENDED SELF-TUNING ALGORITHM [52]

As indicated in section 4.5, the basic self-tuning property hinges on the fact that the noise filter $C(z^{-1})$ can be assumed to be unity during a self-tuning design. However, Clarke's approach implicitly estimates the noise dynamics because $C(z^{-1})$ appears as a factor in $\widetilde{H}(z^{-1})$ (cf. eqn. 4.51).

It seems therefore that a contradiction exists concerning the assumptions which can be made about the noise dynamics. In fact, a closer study of the basic self-tuning property in [3] resolves this contradiction and reveals a deeper result which is known as the extended self-tuning property [52].

The extended self-tuning property tells us that with a system described by equation (4.1) a self-tuning control algorithm can be designed in which $C(z^{-1})$ is modelled by a noise filter which has any desired form. Thus if we model the system as

$$y(t) = - \widetilde{A}(z^{-1}) y(t) + \widetilde{B}(z^{-1}) u(t) + \widetilde{C}(z^{-1}) \xi(t), \qquad (4.54)$$

where $\widetilde{C}(z^{-1})$ is an arbitrary polynomial, the self-tuning system can still converge to the correct closed-loop configuration.

The extended algorithm therefore offers a new degree of freedom in self-tuning design. Moreover, this freedom can be exploited in a number of ways as discussed in [52]. The most important practical consequence however is the extended self-tuning algorithm (ESTA) which combines a regulation and servo-tracking objective within a pole-assignment formulation.

ESTA uses an extended recursive least squares algorithm to estimate the polynomials $\widetilde{A}(z^{-1})$, $\widetilde{B}(z^{-1})$ and $\widetilde{C}(z^{-1})$ in the model

$$y(t) = - \widetilde{A}(z^{-1}) y(t) + \widetilde{B}(z^{-1}) u(t) + \widetilde{C}(z^{-1}) \xi(t). \qquad (4.55)$$

Control is then applied using the composite servo/regulation configuration of Fig. 4.4. Thus

$$u(t) = - F(z^{-1}) u(t) - G(z^{-1}) y(t) + S(z^{-1}) y_r(t) \qquad (4.56)$$

where the controller polynomials $F(z^{-1})$, $G(z^{-1})$ are obtained from the identity

$$(1+F(z^{-1}))(1+A(z^{-1})) + \tilde{B}(z^{-1}) G(z^{-1}) = \tilde{C}(z^{-1}) T(z^{-1}) \qquad (4.57)$$

Where $T(z^{-1})$ represents the desired closed-loop pole set.

If the self-tuning system converges the closed-loop system is

$$y(t) = \frac{S(z^{-1})\tilde{B}(z^{-1})}{T(z^{-1})\tilde{C}(z^{-1})} y_r(t) + \frac{1+F(z^{-1})}{T(z^{-1})} \xi(t) \qquad (4.58)$$

The precompensator $S(z^{-1})$ can take one of many forms. However the simplest is

$$S(z^{-1}) = \frac{\tilde{C}(z^{-1})T(1)}{\tilde{B}(1)} \qquad (4.59)$$

This choice of compensator ensures that $y(t)$ tracks $y_r(t)$ in the steady state, and that the modes of $C(z^{-1})$ are cancelled from the servo response. Other choices of $S(z^{-1})$ are possible to remove some, or all, of the modes associated with $\tilde{B}(z^{-1})$. However, these involve polynomial factorization and cancellation procedures [17] and as such are time consuming and to be avoided because they are sensitive to changes in the open-loop zeros.

Notice that the extended self-tuner (Fig. 4.4) does not incorporate a digital integrator in the forward loop, this is because the steady state tracking property is achieved by the scaling coefficients $\tilde{B}(1)$ and $T(1)$. Attempts to incorporate incremental action can cause problems, because it will introduce a factor $(1-z^{-1})$ in $\tilde{C}(z^{-1})$. This in turn introduces a differentiator into $S(z^{-1})$, with the result that the servo-precompensator will 'switch-off' when $y_r(t)$ is slowly changing.

The extended self-tuner can be seen as a synthesis of the simple regulator and servo-algorithms outlined earlier, and as such, combines the advantages of both. An additional advantage of ESTA lies in the explicit estimation of $\tilde{C}(z^{-1})$, since this allows the self-tuner to be commissioned open-loop in the manner proposed by Clarke [27].

On the other hand there are occasions, and many process control loops are in this category, when the regulator format of Fig. 4.3 is appropriate. In such circumstances the extended self-tuning property can still be deployed to assist in convergence and to formulate implicit algorithms with which we can avoid the need to solve the pole-assignment identities and the pseudo-commutation in multivariable problems (see [52] for a discussion of implicit pole-assignment self-tuning).

4.9. NON-PARAMETRIC SELF-TUNERS

The methods outlined here are aimed at designing DDC loops based upon a z-transform model of an underlying continuous time system. A related procedure is to use non-parametric identification methods [37] and hence identify an unparametrized model which relates to the continuous time system, and design a control loop accordingly. An example of this approach is the non-parametric self-tuning vibration tester described in

[38,48]. Incidentally this form of self-tuner is widely used in the vibration testing area although it is not usually referred to as a self-tuning algorithm. Further discussion on non-parametric self-tuning is given in [51].

4.10. SELF-TUNING PREDICTION [29] (See also Appendix 4.1)

Stochastic regulation and prediction are closely related problems [24]. Indeed as mentioned previously, self-tuning can be set up as a 'k-step prediction' model, which has its output nullified by control action at each time interval (see the proof in Section 3 of [2] and the discussion of Section 4.5). More important than its use in proofs is the fact that prediction can be formulated as a self-tuning problem [29]. This leads to an extremely useful class of algorithms with excellent predictive properties. Their virtue is that the time-consuming aspects of traditional forecasting methods [39] are avoided, and an inherent parameter adaptive capability is obtained (see [40] for an application). By the same token self-tuning prediction has uses in quality testing and health monitoring [46].

4.11. APPLICATIONS

A wide range of self-tuning applications now exist. (See for example [33,34,15,21,41,42,43,40]). For the most part the applications have been to fairly slow systems (often chemical processes) with a requirement for good regulatory performance. As a result there is relatively little experience with fast systems traditionally associated with the servo-mechanism problem. Preliminary experience [15] however indicates that actuator non-linearity (which is always a problem) gives the self-tuner some significant difficulties. Supervisory logic helps here [10] and it may well be that a combination of open-loop supervision and closed-loop adaptation is the best practical approach [42].

Self-tuning has also found significant application in the fringe control topics of prediction [40,43]. Again this links across to the notions of filtering which can be written in a self-tuning format [44]. In this context, self-tuning is of great practical potential in the area of prediction and filtering, not least because it can be used as a forecasting tool. The reader should also recall that reliability engineering draws extensively from statistical forecasting methods, and self-tuning can contribute here too [45].

4.12. FURTHER TOPICS

The discussion here has centred around the basic features of self-tuning. However, as hinted earlier, self-tuning is one form of algorithm drawn from the general class of recursive algorithms [32]. It therefore relates to recursive parameter and state estimators in a manner which has only been hinted at here. By the same token the close link to optimal filtering [44] requires theoretical elaboration.

From a functional viewpoint considerable further discussion (and indeed research) would be required to fully explain the design and execution of a self-tuning experiment. Some guidance is given in [10,31], however this is by no means comprehensive and much remains to be done on the 'software engineering' aspects of self-tuning.

It is also pertinent to note that self-tuning originated in the system identification camp. Accordingly, it has been developed as an identification tool by identification experts, with relatively little input from the control theorists. Thus questions of closed-loop integrity, and general robustness of the early self-tuners were not questioned. The pole-assignment approach provides an answer to the robustness question [16], but much more work is required in this area.

The area of self-tuning is closely related to the techniques of model reference adaptive control (MRAC) and although this area has not been mentioned here there are many ways of indicating and utilising the links between self-tuners and MRAC systems. It is however appropriate to note that many MRAC schemes involve cancellation of system dynamics, and again the integrity of such systems is not assured.

4.13. CONCLUDING REMARK

The classical methods of Nyquist, Bode and the like have found wide use mainly because they offer analysis and design procedures which operate <u>directly</u> upon something which can be easily and accurately measured — the frequency response function. This point is obvious, but essential because for any design tool to find acceptance it must involve a link with a readily identifiable system model. It is the author's belief that self-tuning will prove durable if only because it obeys this precept by linking least squares estimation with a set of controller synthesis rules which operate directly upon the estimated coefficients of the least squares model. Beyond this however, self-tuning is potentially important because it constitutes a movement toward a class of intelligent control systems which combine algorithmic simplicity with an adaptive capability.

In compiling this chapter I have drawn freely from the open literature. The underlying theme (pole-zero assignment) however is the result of a team effort by colleagues and students at the UMIST Control Systems Centre. The work has been jointly funded by the Guilbenkian Foundation, the British Council, the Central Electricity Generating Board and the Science Research Council.

TABLE 1   <u>z-transforms for systems involving fractional time delays</u>

In this table τ is a transport delay such that $0 <= \tau <= T$ and T is the sample interval. Also $z \underline{\underline{\Delta}} \exp(-T s)$ and a zero order hold is assumed to be present.

s-domain	z-domain

$se^{-\tau s}$
$$\frac{T-\tau}{T^2} z^{-1} + \frac{2\tau-T}{T^2} z^{-2} - \frac{\tau}{T^2} z^{-3}$$

$e^{-\tau s}$
$$\frac{(T-\tau)}{T} z^{-1} + \frac{\tau}{T} z^{-2}$$

$\dfrac{e^{-\tau s}}{s}$
$$\frac{(T-\tau) z^{-1} + \tau z^{-2}}{1-z^{-1}}$$

$\dfrac{ae^{-\tau s}}{s+a}$
$$\frac{(1-e^{-aT}e^{a\tau}) z^{-1} + e^{-aT}(e^{a\tau}-1) z^{-2}}{1-e^{-aT}z^{-1}}$$

$\dfrac{e^{-\tau s}}{s^2}$
$$\frac{(T-\tau)^2 z^{-1} + (T^2+2T\tau-2\tau^2) z^{-2} + \tau^2 z^{-3}}{(1-z^{-1})^2}$$

$\dfrac{a^2 e^{-\tau s}}{(s+a)^2}$
$$\frac{[1-e^{-aT}e^{a\tau}(1+aT-a\tau)] z^{-1}+e^{-aT}[-2+e^{a\tau}(1+aT-a\tau)+e^{-aT}e^{a\tau}(1-a\tau)] z^{-2} + e^{-2aT}[1-e^{a\tau}+a\tau e^{a\tau}] z^{-3}}{(1 - e^{-aT}z^{-1})^2}$$

$\dfrac{sa^2 e^{-\tau s}}{(s+a)^2}$
$$\frac{a^2 e^{-aT}e^{a\tau}[(T-\tau) z^{-1} + (\tau e^{-aT}-T+\tau) z^{-2} - \tau e^{-aT}z^{-3}]}{(1-e^{-aT}z^{-1})^2}$$

4.14 REFERENCES

1. ASTROM, K.J. and WITTENMARK, B.: 'On Self-Tuning Regulators', <u>Automatica</u>, 1973, 9, pp.185-199.
2. CLARKE, D.W. and GAWTHROP, P.B.: 'Self-Tuning Controller', <u>Proc. IEE</u>, 1975, 22, pp.929-934.
3. WELLSTEAD, P.E., EDMUNDS, J.M., PRAGER, D.L. and ZANKER, P.: 'Self-Tuning Pole-Zero Assignment Regulators', <u>Int. J. Contr.</u>, 1979, 30, pp.1-26.
4. WELLSTEAD, P.E. and ZANKER, P.: 'Servo-Self Tuners', <u>Int. J. Contr.</u>, 1979, 30, pp.27-35.
5. WELLSTEAD, P.E., PRAGER, D. and ZANKER, P.: 'Pole Assignment Self-Tuning Regulator', <u>Proc. IEE</u>, 1979, 126, pp.781-787.
6. LJUNG, L.: 'Convergence Concepts for Adaptive Structures', Report 7218, Division of Automatic Control, Lund Institute of Technology, Lund, Sweden, 1972.
7. GOODWIN, G.C., RAMADGE, P.J. and CAINES, P.E.: 'Discrete Time

Multivariable Adaptive Control', <u>SIAM Jour. on Cont. and Opt.</u>, (to appear 1979-80).

8. RAGAZZINI, J.R. and FRANKLIN, G.F.: 'Sampled Data Control Systems', (McGraw-Hill, New York, 1958).

9. WELLSTEAD, P.E.: 'Aliasing in System Identification', <u>Int. J. Contr.</u>, 1975, 22, pp.363-375.

10. WELLSTEAD, P.E. and ZANKER, P.M.: 'Techniques of Self-Tuning', Control Systems Centre, Report 432, UMIST, 1979. (Available from Control Systems Centre Secretariat).

11. SEAL, H.: 'Historical Development of the Gauss Linear Model', <u>Biometrika</u>, 1967, 56, pp.1-24.

12. LAWSON, C.L. and HANSON, R.J.: 'Solving Least Squares Problems', (Prentice Hall, 1974).

13. ROBINS, A.J. and WELLSTEAD, P.E.: 'Fast Algorithms for System Identification', Control Systems Centre Report 459, UMIST 1979. (Available from Control Systems Centre Secretariat).

14. YOUNG, P.C.: 'Recursive Approaches to Time-Series Analysis', <u>Bulletin Inst. of Maths. and Applications</u>, 1974, 10, pp.207-224.

15. WELLSTEAD, P.E. and ZANKER, P.: 'Practical Features of Self-Tuning', <u>Proc. IEE Conference Trends in On-line Computer Control Systems</u>, Sheffield, 1979. (Available as C.S.C. Report 461 from the Control Systems Centre Secretariat).

16. WELLSTEAD, P.E., EDMUNDS, J.M., PRAGER, D.L. and ZANKER, P.: 'Classical and Optimal Self-Tuning Control', <u>Int. J. Contr.</u>, 1980, pp.610-612.

17. ASTROM, K.J., WESTERBURG, B. and WITTENMARK, B.: 'Self-Tuning Controllers based on Pole-Zero Placement', Report TRFT 3148, Dept. of Automatic Control, Lund Institute of Technology, 1979.

18. WELLSTEAD, P.E.: 'Instrumental Product Moment Test for Model Order Estimation', <u>Automatica</u>, 1978, 14, pp.89-91.

19. ROBINS, A.J.: M.Sc. Dissertation, Control Systems Centre, UMIST, 1977. (Available from UMIST Library via the Inter-Library Loan Service).

20. ALBERT, A. and GARDNER, L.: 'Stochastic Approximation and Nonlinear Regression', Monograph 42, MIT Press, Cambridge, Mass., 1967.

21. FORTESCUE, T.R., KERSHENBAUM, L.S. and YDSTIC, B.E.: 'Self-Tuning Regulators with Variable Forgetting Factors', Internal Report, Dept. of Chemical Engineering, Imperial College, London, 1979.

22. EDMUNDS, J.M.: Ph.D. Thesis, Control Systems Centre, UMIST, 1976.

23. ZANKER, P.M.: Ph.D. Thesis, Control Systems Centre, UMIST, 1980.

24. ASTROM, K.J.: 'Introduction to Stochastic Control', Academic Press, London, 1970.

25. TRUXAL, J.G.: 'Automatic Feedback Control System Synthesis', (McGraw-Hill, 1955).

26. CLARKE, D.W. and HASTINGS, J.: 'Design of Digital Controllers for Randomly Disturbed Systems', <u>Proc. IEE</u>, 1971, 118, pp.1503-1506.

27. CLARKE, D.W. and GAWTHROP, P.J.: 'Self-Tuning Control', <u>Proc. IEE</u>, 1979, 126, pp.633-640.

28. ALLIDINA, A. and HUGHES, M.J.: 'Generalised Self-Tuning Controller with Pole-Assignment', <u>Proc. IEE (Part D)</u>, 1980, 1, pp.13-18.

29. WITTENMARK, B.: 'A Self-Tuning Predictor', <u>Trans. IEEE</u>, 1974, AC-19, pp.848-851.

30. EYKHOFF, P.: 'System Identification', (Wiley-Interscience, London, 1974).

31. ZANKER, P.M. and WELLSTEAD, P.E.: 'Computer-Aided Self-Tuning Controller Design: A PDP 11 programme suite', Report 435, Control

Systems Centre, UMIST, 1978. (Available from the Control Systems Centre Secretariat).
32. LJUNG, L.: 'Analysis of Recursive Stochastic Algorithms', Trans. IEEE, 1977, AC-22, pp.551-575.
33. BORISSON, U.: 'Self-Tuning Regulators – Applications and Multivariable Theory', Report 7513, Dept. of Automatic Control, Lund Institute of Technology, Lund, Sweden, 1975.
34. KEVICZKY, L. et al.: 'Self-Tuning Adaptive Control of Cement Raw Material Blending', Automatica, 1978, 14.
35. PRAGER, D.L. and WELLSTEAD, P.E.: 'Multivariable Pole Assignment Self-Tuning Regulators', Report 452, Control Systems Centre, UMIST, 1979. (Available from the Control Systems Centre Secretariat).
36. Special Issue on System Identification. (Especially the first three papers), IEEE Trans., 1974, AC-19, Number 6.
37. WELLSTEAD, P.E.: 'Non-Parametric System Identification', Proc. IFAC Symposium on System Identification and Parameter Estimation, Darmstadt, F.R.G., (1979). (Proceedings published by Pergamon Press, London).
38. CARVALHAL, F.J.G.G., WELLSTEAD, P.E. and PERIARA, J.J.: 'Identification and Adaptive Control in the Simulation of Seismic Disturbances', Proceedings of IFAC Symposium on Identification and System Parameter Estimation, Tbilisi, SSRG, 1976. (Available from Control Systems Centre Secretariat as C.S.C. Report 328).
39. BOX, G.E.P. and JENKINS, G.M.: 'Forecasting and Control', (Holden-Day, San Francisco, 1968).
40. WELLSTEAD, P.E., GALE, S. and MONTAGNER, J.: 'Self-Adaptive Prediction of Blast-Furnace Hot Metal Quality', Control Systems Centre Report 403, UMIST, 1979. (Available from the Control Systems Centre Secretariat).
41. MORRIS, A.J., FENTON, T.P. and NAZER, Y.: 'Application of Self-Tuning Regulators to the Control of Chemical Processes', IFAC Congress on Digital Computer Applications to Process Control, (Holland, 1977).
42. ZANKER, P. and WELLSTEAD, P.E.: 'Self-Tuning Diesel Engine Regulation', Report 422, Control Systems Centre, UMIST, 1978. (Available from Control Systems Centre Secretariat).
43. IEE Colloquium Digest No. 1980/21. 'Developments in Self-Tuning Systems', 1980.
44. YOUNG, P.C.: 'Self-Tuning Kalman Filter', Electronics Letters, 1979, 15, pp.358-360.
45. PRAGER, D.L. and WELLSTEAD, P.E.: 'Interactive Maximum Likelihood Estimation', Report 458, Control Systems Centre, UMIST, 1979. (Available from Control Systems Centre Secretariat).
46. FERNANDO DEL BUSTO, R.: M.Sc. Dissertation, Control Systems Centre, 1978. (Available from UMIST Library via Inter-Library loan services).
47. WELLSTEAD, P.E.: 'Scale Models in Control Systems Engineering', Report 482, Control Systems Centre, UMIST, 1980. (Available from the Control Systems Centre Secretariat).
48. CARVALHAL, F.J.G.G.: 'Controle Adaptativo na Simulacao de Solicitacoes Sismicas', Ph.D. Thesis, National Laboratory for Civil Engineering, Portugal, 1979. (Report 16/13/5588, LNEC, Av.do.Brasil, Lisboa, Portugal).
49. PRAGER, D.L.: Ph.D. Thesis, Control Systems Centre, 1980.
50. TRUXAL, J.G.: 'Automatic Control System Synthesis', (McGraw-Hill,

1950).

51. WELLSTEAD, P.E.: 'Self-Tuners – a nonparametric approach', Control Systems Centre Report, UMIST, 1980. (Available from the Control Systems Centre Secretariat).

52. WELLSTEAD, P.E. and SANOFF, S.P.: 'Extended self-tuning algorithm', Int. J. Contr., 1981, 34, pp.434-455.

APPENDIX 4.1

SELF-TUNING PREDICTION AND RECURSIVE APPROXIMATE MAXIMUM LIKELIHOOD

Recall now the remarks concerning the correspondence between self-tuning and recursive parameter estimation. This correspondence can be illustrated by considering the one step ahead prediction of data y(t) created by the system

$$(1+A(z^{-1}))y(t) = (1+C(z^{-1}))\xi(T) \tag{4.60}$$

where $\xi(t)$ is a driving white noise sequence.

The one-step-ahead prediction error is defined as $\varepsilon(t+1)$, thus –

$$\varepsilon(t+1) = y(t+1) - \tilde{y}(t+1|t) \tag{4.61}$$

where $\tilde{y}(t+1|t)$ is the prediction of y(t+1) given data up to time t.

Now, neglect the dependence upon z^{-1} and substitute 4.61 in 4.60 to get

$$\varepsilon(t+1) = -A\varepsilon(t+1) - (1+A)\tilde{y}(t+1|t) + (1+C)\xi(t+1) \tag{4.62}$$

The self-tuning predictor assumes that $C(z^{-1})$ in equation 4.62 is zero and estimates $\tilde{A}(z^{-1})$ and $\tilde{B}(z^{-1})$ in the least squares regression model –

$$\varepsilon(t+1) = -\tilde{A}\varepsilon(t+1) - (1+\tilde{B})\tilde{y}(t+1|t) + w(t+1) \tag{4.63}$$

and constructs a predictions at each step,

$$\tilde{y}(t+1|t) = \frac{\tilde{A}}{1+\tilde{B}} \varepsilon(t+1) \tag{4.64}$$

which sets the first two terms on the right hand side of eqn. 4.63 to zero, such that $\varepsilon(t+1) = w(t+1)$. Here w(t+1) is the least squares fitting error or residual.

Now substituting eqn. 4.61 in eqn. 4.63 and rearranging gives the equivalent expression

$$y(t+1) = -\tilde{B}y(t+1) + (\tilde{B}-\tilde{A})\varepsilon(t+1) + w(t+1) \tag{4.65}$$

But the predictor forces $\varepsilon(t)$ to equal w(t) at each stage. Therefore eqn. 4.65 can be rewritten as

$$y(t+1) = -\tilde{B}y(t+1) + (\tilde{B}-\tilde{A})w(t+1) + w(t+1) \tag{4.65 bis}$$

which is the extended least squares algorithm from which many approximate maximum likelihood methods may be derived. In this way the link between self-tuning and recursive estimation can be seen. Moreover, if the fitting error w(t) converges to the driving noise $\xi(t)$, as required by the self-tuning property

i.e. $w(t) \rightarrow \xi(t)$

Then we have

$$\tilde{B}(z^{-1}) \rightarrow A(z^{-1})$$

and $$\tilde{B}(z^{-1}) - \tilde{A}(z^{-1}) \rightarrow C(z^{-1})$$ (4.66)

Again this indicates the relationship between self-tuning and recursive estimation. In addition the relations 4.60 relate back to remarks about how the system parameters are scrambled up into the self-tuning estimates, by the rearrangement of the system and measurement equations.

APPENDIX 4.2

LIST OF SYMBOLS

General notational points

  i) tild usually indicates either an estimated quantity (e.g. $\tilde{\theta}$) or expected values of least square estimates.

 ii) z^{-1} is variously used as the complex variable e^{-sT} and the backward shift operator.

$A(z^{-1})$, $B(z^{-1})$, $C(z^{-1})$ system polynomials in z^{-1}

$C(z^{-1})$ surrogate noise filter

$D(z^{-1})$ actual closed loop characteristic polynomial

$F(z^{-1})$, $G(z^{-1})$, $H(z^{-1})$, $S(z^{-1})$ controller polynomials in z^{-1}

$\underline{F}(z^{-1})$, $\underline{G}(z^{-1})$, $\underline{\tilde{F}}(z^{-1})$, $\underline{\tilde{G}}(z^{-1})$ polynomial matrices in z^{-1}

$P(z^{-1})$, $Q(z^{-1})$, $R(z^{-1})$, $E(z^{-1})$ polynomials in z^{-1}

$\underline{P}_t$ scaled inverse covariance matrix at time t

$\underline{\bar{P}}_t$ finite memory version of $\underline{P}_t$

$\underline{R}$ positive definite matrix

$T(z^{-1})$ desired closed-loop characteristic polynomial

T controller sampling interval

$W(z^{-1})$		transfer function in z^{-1}
$\underline{Z}(t)$		vector of delayed input/output variables
a_i	$(i=1, \ldots n_a)$	coefficient of z^{-1} in $A(z^{-1})$
b_i	$(i=1, \ldots n_b)$	coefficient of z^{-1} in $B(z^{-1})$
c_i	$(i=1, \ldots n_c)$	coefficient of z^{-1} in $C(z^{-1})$
$\xi(t)$		white noise sequence
f_i	$(i=1, \ldots n_f)$	coefficient of z^{-1} in $F(z^{-1})$
f_c		achievable closed-loop bandwidth
g_i	$(i=1, \ldots n_g)$	coefficient of z^{-1} in $G(z^{-1})$
h_i	$(i=1, \ldots n_h)$	coefficient of z^{-1} in $H(z^{-1})$
k		discrete system time delay
s_i	$(i=1, \ldots n_s)$	coefficient of z^{-1} in $S(z^{-1})$
t	$(=0, \pm1, \pm2, \ldots)$	time index
t_i	$(i=1, \ldots n_t)$	coefficient of z^{-1} in $T(z^{-1})$
$u(t)$		sequence of controller output variables
$y(t)$		sequence of system output variables
$y_r(t)$		reference value or desired value of output
$\Delta u(t) = u(t) - u(t-1)$		the control increment
ξ		damping factor of second order system
$\varepsilon(t)$		moving average process
$\xi'(t)$		least squares fitting error
$\emptyset(t)$		pseudo system output
$\underline{\theta}(t)$		vector of system polynomial coefficients
ρ		forgetting factor
ω_n		natural frequency of second order system

Chapter 5

Requirements for real-time computing

5.1 INTRODUCTION

A process control computer system consists essentially of a central processor together with a variety of standard or special equipment which allows it to communicate both directly with plant and with plant personnel. This communication must be both effective and efficient, with the processor capable of sufficiently rapid execution to provide real-time control action.

Such considerations as the compatibility of the input and output channels of the system with the various signal-generating devices and process instruments, and the degree of control and flexibilty which the computer has in scanning the various input and output devices, are consequently important. Computer characteristics including storage capacity, speed of operations, flexibility, command and word structure, and programming features will influence the speed of execution.

Computer control systems are normally designed on a modular basis and can be assembled and tailored economically to satisfy specific process requirements. Basic to each system, however, and determining its capabilities are the interrelationships and characteristics of the input and output, processing, storage, and priority-interrupt features. A simplified block diagram of a control computer system is shown in Fig. 5.1. The primary module is the central processing unit (CPU) which, through programmes stored internally in a working storage, directs the operational procedures of the entire system. Closely associated with the central processing unit are a large-volume bulk-storage device for data and programmes, input and output equipment to permit programme loading and information output, and a computer console for manually intervening and controlling the processor. Together these modules represent a typical general-purpose digital computer. Of equal importance to the control computer system, however, are the input and output channels which allow the central processor to communicate directly with the process instrumentation; these are shown to the left of the central processing unit in Fig. 5.1 and indicate the functional modules needed to interface to the various signals which are generated within the process. Of course, the computer system may also be connected to other computer systems, and in large applications may form part of a hierarchy of control computers.

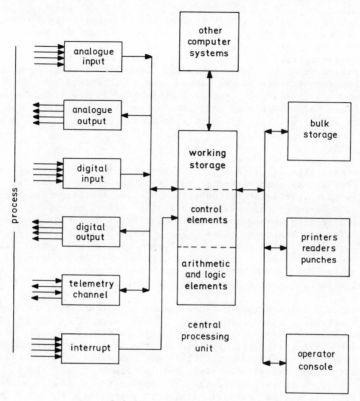

Fig. 5.1 A control computer system

5.1.1 Central Processing Unit

The CPU provides the control and arithmetic capabilities of the system
and takes its instructions from programmes stored in its working
storage. Important features of CPU design include word size,
instruction-set and addressing methods, information transfer, and
priority interrupt. These influence programming, effective computing
speeds, and storage utilization.

The size and structure of words must be such as to ensure adequate
precision in calculations and preferably to allow direct addressing of
any storage location within one instruction word. The command lists of
process control computers should be designed to facilitate process
programming. Desirable features include commands to transfer
variable-length blocks of data between storage units or locations within
storage, single commands which carry out more than one operation,
flexible addressing modes for direct and immediate addressing, and
address modification by use of index registers. These features reduce
the number of 'housekeeping' instructions, thereby benefiting the system
in two ways: they save storage, and they improve the overall processing
speed by reducing the number of storage-access cycles required to

service such instructions.

Information transfer within the processor, between the processor and backing store, and with input and output channels and devices, strongly influences the primary capabilities of the overall control computer system. Because in an on-line application the central processing unit is often called upon to communicate with several devices at once and also to perform computations at the same time, it should be capable of simultaneous or essentially parallel operation. In addition, for maximum flexibility and efficiency of the system, a multilevel interrupt structure should be built into the processor.

5.1.2 Storage

The storage capacity of a control computer system consists of two main types: fast access memory storage and auxiliary storage. The fast access memory is that section of the central processor's storage which contains the data, programmes and results currently being processed, together with the operating system and service routines. It permits random access to the information stored there, short-access times, parallel-by-bit operation, and flexibility in communicating with many external asynchronous input and output devices and auxiliary storage units. Additionally ROM, PROM or EPROM memory may be provided for storage of critical code or for predefined standard functions.

Typically, a large volume of information and programmes is necessary for a control computer application, but most of it is not required simultaneously. Therfore, instead of using additional memory with its relatively high cost, magnetic-disc or drum storage devices may be used as bulk auxiliary storage units. These are currently less expensive per byte, but have relatively long access times. Since they operate asynchronously with respect to the CPU, they require added equipment to accomplish information transfer between them and the CPU.

Protected storage is essential to ensure maximum availability of the system. Memory should be protected against loss of programmes due to power failure and against interference from other programmes especially those being debugged. Protection is also possible by using memory mapping techniques that inhibit any instructions which attempt to store into a protected portion of storage.

Expandable memory and auxiliary storage is normally a requirement for a control computer because of the probability of continual growth of a given computer control application. Memory is usually available in modules of from 4 kbytes to 64 kbytes, expandable to several megabytes. Disc storage offers the most economical bulk-storage medium while retaining reasonable access times, with capacaties in excess of 200 mbytes readily available.

5.1.3 Input and Output

Three types of information transfer are possible namely programmed entry, buffered entry, and direct entry to working storage.

(a) Programmed entry. In programmed entry, every piece of information into or out of the processor's storage proceeds under direct control of programme commands. Input and output cannot therefore proceed simultaneously or be overlapped with computations. If only this method of communication were available, the real-time capabilities of the system would be extremely poor. Consequently control computer systems employ extensive buffering.

(b) <u>Buffering entry</u>. Buffers are storage registers where information being transferred is temporarily held until the processor is ready to receive the information or the peripheral devices and channels have time to operate upon it. They eliminate timing or synchronization problems between the processor and peripheral devices and allow the processor to proceed with computations after initiating a command to the peripheral units. The processor is only informed on completion of the transfer.

(c) <u>Direct entry</u>. Direct input to or output from storage is the most straightforward method for peripheral units to communicate with the central processing unit. This method of entry is known as direct memory access (DMA) or sometimes as 'cycle stealing' because, in essence, the input and output device uses a storage read or write cycle without otherwise affecting the operation of the main processor. Direct entry techniques unburden the processor, allow essentially simultaneous computations with input/output functions, and greatly improve real-time characteristics. Their principal disadvantages are the requirement for external control circuitry and the additional cost.

The control computer system has input and output communication channels which are the means through which the system communicates with the external world, the process, and the human operators. The number, variety, and flexibility of these sections provide the system with its on-line capabilities.

There are three main input and output (I/O) sections, process I/O, operator I/O, and computer I/O. The process I/O can further be divided into analogue and digital subsections. The process analogue channel provides the interfacet that translates the continuous signals associated with the process into a sampled data form for the computer, and vice versa. The process digital channel performs the necessary translation to and from process status (on-off) devices and digital sensors. The operator's channel, or console, provides the translation and communication links to the human operators of the process. The computer I/O channel is: the basic programme-loading channel, the output channel for machine-readable information, and the communication link to programmers and service-engineering personnel for diagnostic input and output.

Common characteristics of the I/O channels derive from: the asynchronous operation of the peripheral devices (with respect to the processor), the operating-speed differences between device and processor, and the desire for flexible information transfer. Most external devices operate at very slow speeds compared with the computer and would severely limit the speed of the control computer system if directly controlled by the processor. Yet maximum flexibility, through internal programme control, is required in the selection, translation, and control of input or output on a point-by-point basis. These requirements lead to designs where all three forms of information transfer might be provided in the system, namely programmed entry, buffered entry, and direct storage entry.

Increasingly intelligent I/O processing is being incorporated into the design of control computers to unburden the processor of highly repetitive functions. One such function might be the comparison of a measurement with a standard value to detect randomly and infrequently occuring events such as out-of-limit inputs, or status changes. Modern control computer systems employ highly buffered and/or direct-entry I/O channels which incorporate some basic computing functions, such as auto-driver channels. These features simplify programming, reduce interrupt operations, provide more efficient storage utilization, and improve the overall capabilities of the control computer system.

Buffer registers are used extensively with the I/O channels. These registers temporarily hold and transfer the data, while still other registers store instructions for controlling the channels and their peripheral devices. The number and size of these buffers vary, but the minimum requirement for a control system is a one-character or one-word data register and an I/O addressing and control register. In this manner, input/output transfer and computer processing can occur essentially simultaneously. Buffer registers for each type of I/O channel make possible overlapping of processing, analogue I/O, digital I/O, and logging.

Modularity of the I/O sections is essential in meeting the multiplicity of process requirements. Control computers must be easily and economically adaptable to a great variety of signals, environments, and applications. An important requirement is that the capabilities and capacity for all input and output functions be inherent in the basic design of the computer system.

5.2 INTERRUPTS

One of the most difficult problems in the design of I/O sections is timing. The CPU must determine when a peripheral has new data or is ready to accept data. Slow peripherals such as switches and indicator lights, which do not require rapid response, cause no difficulties, the CPU may transfer data to or from such devices at any time; the only problem is unresponsiveness if the processor waits too long.

Faster peripherals are another matter. These devices are generally neither fast enough to keep up with the computer nor slow enough so that any treatment will suffice. Keyboards, teletypewriters, printers, cassettes, modems, data acquisition systems, and many other peripherals fall in this category. As noted earlier, such devices may transfer data either asynchronously or synchronously. If the transfer is asynchronous, the CPU must be sure that the device is ready for each transfer. If synchronous, the CPU must start the process and provide the proper timing.

Waiting for signals and providing time intervals in software waste CPU time and increase the size and complexity of programmes. For instance, if the CPU is waiting for the start of transmission from a device with a maximum transfer rate of ten characters per second, it will find the data ready flag active once every tenth of a second at most. Typically this check could be implemented in three instructions each of which may take 5 μs, the CPU will check the flag approximately 6700 times in a tenth of a second. Clearly a search that is unsuccessful 99.9% of the time is inefficient. Even several such checks will hardly occupy the CPU; for example, the CPU could check ten flags 670 times each tenth of a second. Furthermore, there are additional problems which must be considered. For example what happens if one peripheral has data ready while the CPU is servicing another? Or alternatively what if the first peripheral that the programme checks is almost always active? Checking each peripheral to see if it is ready to transfer data is called polling. This procedure is comparable, in systems of any size or complexity, to handling a telephone switchboard by picking up each line successively to see if a caller is on it.

Synchronous peripherals require a precise time interval between transfers. The CPU can provide the delay itself by loading a register and decrementing it a specified number of times. However, the delay completely occupies the processor. If the CPU is to perform other tasks during the interval, then either the programmer must keep track of the time used or the processor must check a timer to see if the interval is over. Both options greatly increase the complexity of the programme.

The most common alternative to polling and timing programmes is the use of interrupts. An interrupt is an input to the CPU that can directly alter the sequence of operations at the hardware level. The interrupts acts like a buzzer, causing the processor to halt its normal operations and respond to the input. Interrupts are obviously useful for handling input/output since it is necessary for the CPU to check flags or provide timing intervals. Other uses of interrupts include:

* Alarm inputs: sensors, switches, or comparators may provide these inputs, alarm conditions are uncommon, but the response time may need to be very short

* Power fail warning: a power-failure interrupt allows the system to save data in a low-power memory or switch to a backup power supply. (The power-fail-detection circuitry is typically an RC network that senses the loss of power at an early enough stage so that the CPU is able to execute many instructions correctly before the power is completely lost. Power failures are infrequent events that require immediate action: a power-fail interrupt must take precedence over all other activities, since main power failure will cause a complete system shutdown)

* Control panel or manual override: an interrupt can allow external control of a system for field maintenance, repair, testing and debugging

* Debugging aids: interrupts can allow the insertion of corrections, breakpoints or traces

* Hardware failure indicators

* Transmission error indicators

* Coordination of multiprocessor systems

* Control for direct memory access

* Control for operating systems

* Performance measurement

* Real-time clock: the real-time clock simply provides regularly spaced interrupts at specified intervals of time.

An important factor in determining the usefulness of interrupts is the ratio of the required response time to the time between events. If the response time is much shorter than the average time between events, the processor must check the status flag many times. Several such bit-checking operations could seriously diminish processor throughput.

The major disadvantage of interrupts is their random nature. Although this is a key to the usefulness of interrupts, it makes interrupt-driven programmes difficult to debug and test. Interrupts are contrary to the modern trend toward simpler and more carefully defined programmes as represented by structured programming and top-down design. An interrupt-driven programme is actually far less structured than a programme with many specific transfers of control. There are, in fact, potential transfers of control everywhere in the interrupt-driven programme that do not even appear in the listing. Clearly the potential for havoc is enormous.

The usual way to discover errors in interrupt driven software is to run problems that are time dependent. Errors that occur irregularly are usually in the interrupt system, since other parts of the programme can be repeated. Obviously this approach is not scientific; the debugging and testing of any but the simplest interrupt systems is complex and uncertain. Few systems can generate interrupts randomly or properly analyze test results.

Interrupt service routines are often quite difficult to write, since they must operate properly regardless of when the interrupt occurs. The routines may need to save the contents of registers, flags, and memory locations and restore them before returning. Satisfying all the possible requirements may take a large number of instructions, many of which will be wasted in the most common situations. Interrupt systems also require extra hardware, particularly if many sources can cause interrupts. The amount of this hardware increases as the number of sources increases, since rapid identification of the source is essential.

Interrupts are less useful when the I/O data rates are high since they still require data transfers through the CPU. In fact, interrupts introduce additional overheads of their own. At high data rates, polling is less inefficient and timing intervals are easier to generate directly from the processor clock. The problem at data rates exceeding 10 kilobits per second is for the processor to keep up with the data transfers while still doing some useful work. Direct memory access systems, which substitute hardware for software control and provide a direct path between the memory and I/O sections, can greatly increase I/O capability at high data rates, but involve complex hardware.

5.3 <u>CHARACTERISTICS OF INTERRUPT SYSTEMS</u>

All interrupt systems must deal with such basic problems as:

* When are the interrupt inputs examined and what signal characteristics are required?

* How does the processor transfer control to the interrupt service routines?

* How does the processor save the current state of the computer or machine status and restore it after completing the service routine?

* How does the processor determine which source caused the interrupt?

* How can the processor distinguish between high-priority interrupts, such as power failure or alarms, and low-priority interrupts, such as a printer that is ready for more data?

* How can the interrupt system be disabled during programmes that should not be interrupted?

These problems are dealt with below in order.

5.3.1 <u>Interrupt Inputs</u>

Although interrupt systems vary widely, most CPUs examine the interrupt inputs only at the end of instruction cycles. Resuming instruction cycles in the middle would require the saving of many intermediate results. The standard technique, therefore is the one flowcharted in Fig. 5.2. The interrupt signal must be latched, since a complete instruction cycle may be quite long. The signal may also need to be synchronized with the processor clock to ensure recognition.

The number of interrupt input varies. A single input can, of course, have more than one source, since an OR gate can be used to combine the signals.

5.3.2 <u>Interrupt Response</u>

The way in which the CPU responds to an interrupt can also vary considerably. The most popular techniques are:-

(a) executing a CALL or TRAP instruction to a specified address
(b) fetching a new value for the programme counter from a specified register or memory location
(c) executing a CALL instruction to an externally supplied address
(d) using an output signal, INTERRUPT ACKNOWLEDGE, to gate an instruction onto the data bus.

The basic trade-off is between hardware and software. Methods like (a) and (b) require little external hardware but provide no way to identify a source directly. Methods like (c) and (d) are more flexible and can identify sources but require more external hardware.

The processor can transfer control to an interrupt service routine and then back to the original programme in the same way that it transfers control to and from subroutines. Any of the techniques for handling subroutines can be used. The CALL instruction that places the return address in a stack is the most flexible. However, this technique requires external RAM or an on-chip stack, either of which must be share carefully between subroutines and interrupt service routines.

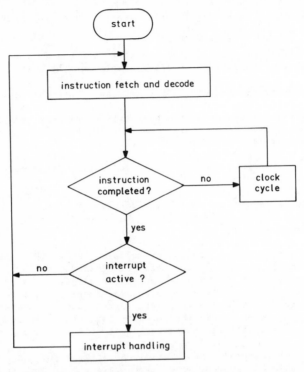

Fig. 5.2 Flowchart of instruction cycle with interrupt examination

Supplying an address or instruction with external hardware can cause many timing and control problems. The external hardware must, of course, not interfere with normal memory cycles; on the other hand, the memory must not interfere with the external hardware when it takes control of the bus. Clearly a one-word address or instruction will be much simpler to produce and control than a multiword address or instruction. TTL or MOS encoders can provide some or all the bits. However, singleword addresses or instructions limit the number of distinct inputs and may interfere with page-zero addressing. The alternative is a collection of registers and timing circuits that can place a multiword instruction on the data bus; special LSI controllers may contain the required hardware.

5.3.3 Saving and Restoring Registers

Most CPUs automatically save the contents of some of the working registers as part of the response to the interrupt input. All CPUs save the old value of the programme counter, but some save additional registers. For example the contents of all the working registers may be stored, a procedure which is convenient and saves time when the registers contain useful information; it wastes time and memory when they do not.

Most processors require several instructions to save the register contents. Three methods are commonly used.

(a) Storing the contents of registers and flags directly in the memory. The resulting interrupt service routine cannot itself be interrupted unless a separate area of memory is available for storing the next level of register contents.

(b) Storing the register contents in a memory stack. This method is simple because the stack pointer contains the storage address; it allows multilevel interrupts. The major inconvenience is that the stack may overflow. The return address is also in the stack, and special exits may require many stack operations.

(c) Switching between sets of registers. A few processors have duplicate sets of registers. The interrupt service routine can simply use the other set. This method is faster than either of the previous two methods but means that some of the processor registers are not always available. Switching does not allow multilevel interrupts, since only two sets of registers exist. An alternative approach is to use a designated area of memory as registers, and so all that the interrupt must do is change the pointer that contains the starting address of the area.

Machine status must, of course, be restored before the interrupt service routine ends. The stack method requires that data be restored in the opposite order from which it was saved. The other methods have straightforward restoring techniques.

The number of registers that must be saved depends on the number used by the main programme and the interrupt service routine. If the main programme simply waits for the interrupt, no registers need be saved or restored. Nor does the interrupt service routine need to save registers that it does not use. In general, interrupt service routines on processors that have only a few registers must save and restore everything; however, this procedure can be accomplished easily and quickly. With processors which have many on-chip registers, the programmer must carefully select the registers to be saved. Otherwise the response time for interrupt may become very long.

5.3.4 Determining Interrupt Sources

If a system has more than one potential source of interrupts, the processor must identify the actual source. Processors with several interrupt inputs can respond differently to each input – that is, transfer control to a different register or memory location. Nevertheless, recognition becomes a problem as soon as the number of sources exceeds the number of inputs.

Two common methods for identifying interrupt sources exist: polling and vectoring. Polling is similar to the normal examination of a data-ready bit; the CPU checks each interrupt bit until it finds one that is active. Vectoring means that each interrupt source provides data (i.e. a vector) that the CPU can use for identification. Vectoring is faster and requires less software; polling, on the other hand, requires less hardware. The advantage of a polling interrupt system over a normal polling system is that the CPU knows that in the interrupt case at least one input is active. The only hardware required is an addressable flip-flop for each interrupt bit.

Clearly a polling interrupt system is only adequate for a small number of sources. Otherwise the time spent identifying the source becomes substantial. Clever software can reduce the average time somewhat by checking the most frequent sources first and by examining groups of sources at a time. Rotation of the order in which souces are checked can keep the average waiting time for all sources the same and can keep one source from blocking the others.

Additional improvements require more hardware. One popular method is the daisy chain in which an acknowledge signal propagates through the sources until it is blocked by the actual source. The daisy chain requires no polling and only a few extra gates; additions or deletions are simple. The acknowledge and enable signals may require extra ports. The constraints on the daisy chain are that the programme cannot change the priorities, and early sources in the chain will block later ones.

The daisy chain is a type of vectoring, since the interrupt source identifies itself. Direct vectoring systems produce the vector with encoders and control signals and place it on the data bus. Fig. 5.3 shows a block diagram of a typical vectored system.

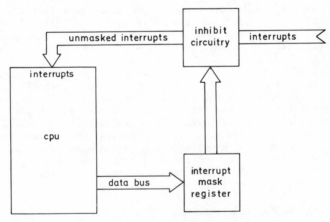

Fig. 5.3 Hardware vectored interrupt structure

The number of different vectors depends on the complexity of the hardware. Producing a large number of vectors requires comlex circuitry, such as a series of encoders or a large PROM. A large interrupt system can employ both vectoring and polling. The vectoring divides the interrupt sources into small groups; polling can then quickly identify a particular source from a group. This combined approach may be much cheaper than complete vectoring and may not require much additional time. Fig. 5.4 shows a software decoded interrupt structure while Fig. 5.5 illustrates the software sequence required.

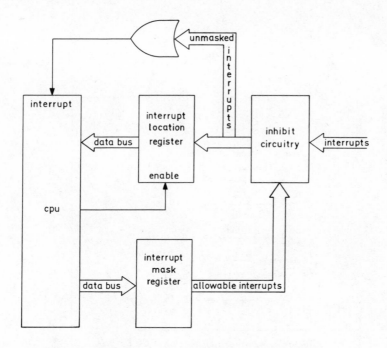

Fig. 5.4 Software decode interrupt structure (i)

The term vectoring refers to the common use of the identification code to produce a jump to a particular service routine. The processor may automatically force the required operation code into the instruction register or the external hardware may have to produce the operation code as well as the vector. Still another alternative is software that forms an address from the identification code and then jumps to that address. Indirect or indexed addressing can provide the jump instruction. This method is slower than the hardware techniques and often results in programmes that are quite difficult to follow. Some processors must manipulate the identification code to form an address, since the code may be only a few bits long. A starting address for the interrupt vectors can easily be introduced into any of these procedures.

5.3.5 Priority

Priority methods involve several questions.

(a) Which of several simultaneous interrupts will the processor service first?
(b) Which interrupts will interrupt other service routines?
(c) How will interrupted service routines be handled?
(d) How will interrupts that are ignored because of low priority eventually be serviced?

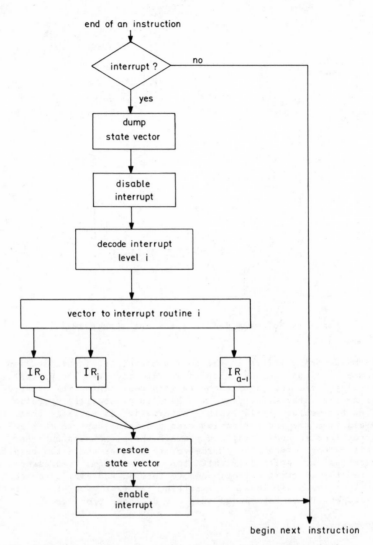

Fig. 5.5 Software decode interrupt structure (ii)

Processors with several interrupt inputs may assign a priority to each input. The highest-priority interrupt that is received at a particular time will be the one that is accepted. Other interrupts will have no effect. Processors having only a single interrupt input can use an external priority encoder to assign priority. TTL encoders will provide a vector and block simultaneous inputs at lower levels. MOS encoders or PROMs can also provide automatic hardware priority.

The priority level may determine which interrupts will be permitted. A processor with several interrupt inputs can simply have an enable or disable associated with each input. This method is the most flexible because it allows priority levels to be individually enabled or disabled. Usually the enabling bits are saved in a register, and so the programmer can determine all the values at once. Internal hardware may set some of the bits automatically when an interrupt is accepted at a particular level.

Processors with a single interrupt will require external hardware to enable or disable different priority levels. The CPU will only respond to those interrupts whose levels have been set to one in the register. The programme must initialize the register to all ones as part of the startup process; the register contents will also have to be stored in memory (since the register is write-only) so that it can be saved and restored during interrupt service routines.

Another option is to exclude all interrupts at priority levels less than or equal to the level of the one that has just been accepted. The equal priority exclusion stops an interrupt from interrupting its own service routine. The external status register must also have a priority disable bit to bypass the comparison process so that the lowest level of interrupts can be accepted (a priority of zero will only allow interrupts at levels greater than but not equal to zero). Fig. 5.6 illustrates the transfer of control during a multilevel interrupt sequence.

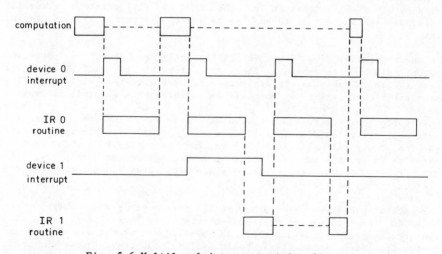

Fig. 5.6 Multilevel interrupt timing diagram

Polling and daisy chaining automatically provide priority. Polling gives higher priority to the inputs that are examined earlier. Daisy chaining assigns priority to the earliest elements in the chain; enabling bits can easily enable or disable all interrupts beyond a particular point.

Interrupts that are ignored because of low priority may cause problems in some systems. Latches, of course, must hold the interrupt signals. However, the service times may become quite long and some interrupts may not be serviced at all. Some helpful techniques include automatically raising the priority of an interrupt each time it is passed over, disabling all or part of the priority system at certain times, or looking for lower-priority interrupts before accepting another high-priority one. Fortunately, few systems are sufficiently complex to require any of these methods.

5.3.6 Enabling and Disabling Interrupts

We have discussed enabling and disabling (or arming and disarming) of interrupts at various times in this section. Almost all CPUs automatically disable interrupts in certain situations. These situations include:

(a) reset; disabling the interrupt system on reset allows the programme to load internal and external registers and initialize variables that may be needed during interrupt service and recognition

(b) after an interrupt has been accepted; disabling the interrupts at this time allows the identification of the interrupt source, the saving of registers, and handling or priority to proceed without further interruption; the disabling also keeps the interrupt from interrupting its own service routine.

Note that interrupts are disabled until specifically enabled and that, as part of the restoration of the CPU to the previous state, the interrupt system must be re-enabled before the interrupt service routine ends.

Some processors have a non-maskable interrupt that the processor cannot disable internally. Such an interrupt is useful for power failure, which obviously takes precedence over all other activities. The non-maskable interrupt can normally be disabled with external hardware.

Interrupt service routines always have some dead time during which the interrupt system is disabled and data may be lost. If interrupts occur so frequently that this is a problem, the system should probably not be based on interrupts at all. Interrupt-driven systems work well only when the processor can handle the maximum data rates and interrupts occur many instructiion cycles apart.

Reentrant programmes are important in interrupt-driven systems. Not only do they allow multilevel interrupts but they also allow service routines to use subroutines that may have been in use at the time of the interrupt. Such subroutines could include code conversions, character manipulation routines, error checking and correction, and I/O handlers. Writing interrupt-driven programmes is difficult enough without having to worry about the hazards of subroutines that may not be reentrant. Fortunately, a real-time operating system will remove much of the burden of handling interrupts from the programmer and will itself contain the most important functions in the form of reentrant code.

5.4 OPERATING SYSTEMS

The operating system is becoming a most important part of the software complex that accompanies a computer system, and the development of operating systems is a major investment on the part of the computer manufacturers. It is a major component of most medium to large computer systems and except for small microprocessor systems it is very difficult to use the computer without some sort of operating system. An inefficient system can have a dramatic effect on the throughput of a computer often nullifying the effect of expensive, fast hardware. Also, since it provides the interface between the user and the computer, any deficiencies in the design of the operating system are constantly brought to the attention of the programmer.

Operating systems have developed in response to a need to increase the utilization of the central processor and peripheral devices and it is still one of the prime functions of an operating system to optimize hardware utilization by automating the flow of work and bringing decisions on management of system resources on to the time-scale of the computer, rather than that of the human operator. However, with the increasing complexity of modern computers operating systems also provide assistance to the user, for example in organizing disc storage as a logical file structure. Also, there is a trend to have some functions, particularly those concerned with input and output, carried out partly by hardware and partly by software, (i.e. firmware) thus adding further functions to the operating system.

5.4.1 Operating system components

Every computer manufacturer tends to produce an operating system tailored to suit a particular range of machines. There are consequently numerous different operating systems in common use; however there are many fundamental features which must be included in all but the most basic systems. A typical system might be divided into a number of major groups which are described below.

(a) **System Manager (Command Processor)** - The System Manager handles all interactions between the system and the console device and provides the operator interface to the operating system. It also contains routines to support a high level command macroprocessor often referred to as a Command Substitution System (CSS), to do memory allocating, and to support direct-access devices. The System Manager controls all I/O requests to the Console and Log devices, and also accepts operator commands from the system console device. It contains logic which provides the console operator with informative messages in case of error. An integral part of the System Manager is often a Command Substitution System (CSS) which builds and executes files of operator commands. A CSS has routines which execute high level macros specified by a single operator command and includes a command parameter substitution facility.
The System Manager should also provide the operator with the command functions necessary to allocate and delete files, display files, rewind, and backspace files assigned to user tasks. It also contains commands used when mounting and dismounting direct access volumes. These functions are executed via appropriate supervisor calls.
Provision should also be made for commands to be entered and executed while user tasks are active, even if tasks have assigned the console device.

(b) <u>Executive Functions</u> – The Executive contains routines to handle supervisor calls, including: miscellaneous utility calls, end-of-task processing, overlay calls, intertask communications and control calls, task status and location swap calls, and task handled supervisor calls.

(c) <u>Task Manager</u> – The Task Manager handles task scheduling functions. A task is usually controlled through a Task Control Block (TCB). At system generation (SYSGEN) time, the user determines the maximum number of tasks the system being built may input. The Task Manager determines when a task Roll out/Roll in is required and handles the Roll process. The user may have typically more than 250 tasks in a system.

(d) <u>Timer Manager</u> – A line frequency clock and a precision interval timer are often used to provide user tasks with a flexible set of timer management/maintenance services. The following services are usually provided: Time-of-day clock, day and year calender, interval and time-of-day wait, interval and time-of-day trap, and driver time-out. Timer trap functions may be set up to occur periodically.

(e) <u>Memory Manager</u> – The Memory Manager handles allocation and freeing of system space, task memory space, and Task Common memory space.

(f) <u>File Manager</u> – The File Manager contains the File Management handler and the I/O interrupt requests to Contiguous, Chained and Indexed Files.

(g) <u>I/O Subsystem</u> – The I/O subsystem consists of an I/O Procedure Handler, the Peripheral Device Drivers, and other routines, such as the System Queue Handler.
The I/O Subsystem provides those system routines and control blocks necessary for device independent I/O requests.

(h) <u>Resident Loader</u> – Resident Loader loads tasks, overlays, library or task common segments. The loader is normally reentrant so that parallel loading is possible.

(i) <u>Floating-Point Support</u> – This support is usually included as a user selected option at system generation time. The support includes storage and restoration of the floating-point registers between task switches. Additionally, if the processor does not have hardware floating-point (either single or double precision), software emulation routines may be included at system generation time to allow for the execution of programmes which use floating-point instructions.

(j) <u>Multi-terminal Monitor</u> – A Multi-terminal Monitor (MTM) provides programme development capabilities at multiple terminals. An MTM normally runs as a privileged task. It provides each terminal with a command structure similar to the command structure of the main console, including batch stream and CSS support to each terminal. Also, MTM can collect job accounting data.
Multiple Background Batch Streams are often also supported in systems containing the Multi-Terminal Monitor, providing support for several background tasks to execute concurrently.

(k) <u>Spooler</u> – Spooling Facilities are normally included within an operating system as a system generation option. The spooling facilities include additional operator commands to control the operation of the spooler. Spooling is accomplished by tasks provided with the system. Input spooling provides for copying batch streams of cards onto disc files for subsequent processing. Output spooling provides for many tasks to share simultaneously one or more print devices.

5.4.2 Overview

Fig. 5.7 shows the principal interactions between the major groupings of an operating system, including foreground and terminal tasks, Resident Library segments, and Task Common segments. For clarity, many minor interactions between these module groupings are not shown.

5.4.3 Supervisor Calls

Supervisor Call instructions provide the programme interface to the operating system. SVC instructions are executed by programmes to request operating system services. Parameters associated with the request are passed to the OS in a user-supplied parameter block. Most of the System Manager's services are performed with SVC instructions, thus making these services available to user tasks.

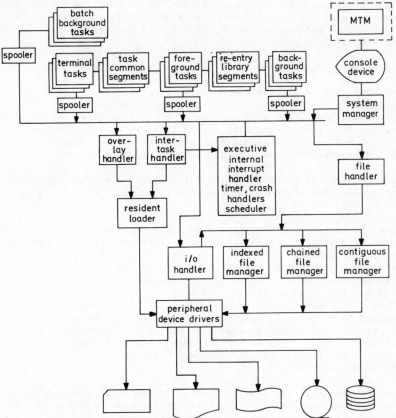

Fig. 5.7 Block diagram of a typical operating system

5.4.4 <u>Memory Management</u>

The memory management portion of an operating system handles
allocation and freeing of memory. Memory is usually divided into two
classes: local memory and global memory. Local memory is that area
containing the OS, all user tasks, System Space, Reentrant Library
segments, and local Task Common segments. Local memory is always
contiguous. All other memory locations are referred to as global or
shared memory. Global memory may be physically contiguous to local
memory or may be located in shared memory banks. In multiprocessor
systems, each system addresses local memory and one or more shared
memory banks. The shared Memory may be used only for Global Task Common
segments.

An operating system normally dynamically manages local memory,
providing allocation and deallocation of memory as needed. Dynamic
memory management allocates memory more efficiently than a partitioned
system, thus minimizing waste. Memory is usually allocated on a
first-fit basis.

(a) <u>Programme Segmentation</u>

Frequently operating systems support three sharable segment types of
programme: Pure, Library, and Task Common, and one non-sharable type,
the Impure segment. The main segment of a task's code is the Impure
segment. The pure segment is the sharable portion of the task's code.
Each task may contain one Impure segment, one pure segment, multiple
Task Common segments, and multiple Library segments.

An assembler can generate programmes segmented into an Impure and Pure
segment. The Pure segment can be shared by multiple copies of the same
task.

When a task is loaded, the loader information block of the task is
read to determine if the sharable segments are currently in memory. If
they are, linkages to these segments are established. If the pure
segment is not in memory, the Pure and Impure are loaded in a contiguous
memory segment. If a required library segment or Task common segment is
not in memory, an error status or message is generated. A required
Library or Task Common segment must have previously been loaded or
allocated by the console operator. All sharable segments are resident.
Task Common and Library segments may be loaded and deleted only via
operator command. A load is not performed unless sufficient memory is
available to satisfy all segments. Pure and library segments are
normally write protected.

Task Common areas are sharable data segments and are normally in
Executive-Protected memory. Therefore, the segments are writable and may
contain data; but, they may not contain executable code. Usually only
foreground tasks can reference Task Common areas.

User tasks can also be arranged to share reentrant library segments.
It is often the responsibility of the console operator to load and
delete libraries from memory. Foreground and Background (Programme
development) tasks may use reentrant libraries. Calls to library
routines are resolved when the task is established.

(b) Roll In/Roll Out (Roll)

To provide memory distribution on a priority basis, operating systems often contain a Roll facility. Roll provides the following benefits:

* The number of tasks within the system is not limited by the amount of memory available.

* Tasks are Rolled in priority order, thus ensuring the greatest mix of high priority tasks at any time.

A task, currently in memory, is rolled out to a disc file when a higher priority task load request requires its memory space. As soon as that task becomes the highest priority rolled task and sufficient memory becomes available to accommodate it, it is rolled back into memory. Roll is commonly used in real-time applications to allow establishment of a low-priority queue (i.e., the rolled task queue) and a high-priority queue (i.e., tasks currently residing in memory). The high-priority real-time tasks remain in control of memory most of the time. When one of these tasks relinquishes its memory, a lower priority task is rolled into memory. This lower priority task is allowed to execute as long as another higher priority task doesnot require its memory. In this way, maximum usage of memory is provided; high and low-priority tasks are running concurrently within the system and a low-priority task is not preventing a high-priority task from executing.

A task is eligible for Roll if:

* it is established to be rollable
* it dynamically sets itself as rollable via a supervisor call
* it is made rollable by another task.

If a higher priority task needs memory space then the following major considerations apply in selecting a lower priority task as a Roll candidate, if:

* it is in an active (ready to run) state
* it is in 'Terminal Wait'
* it is in 'Time-of-Day' wait
* it is suspended
* it is paused
* it is dormant.

A task with process I/O active cannot be eligible for roll and neither can a task with any traps other than 'time of day' active.

A Task Load Supervisor Call can be arranged to cause a roll; consequently, the load operation has two options: Load and Proceed and Load and Wait. When using Load and Proceed, control immediately returns to the task performing the load. The load is accomplished asynchronously. When using Load and Wait, control does not return to the requesting task until the load is complete. If a task performing a load is a lower priority than the task it loaded, the loading task is not rollable until the load operation completes. Specifically, a task cannot normally cause itself to be rolled to accomodate a load it is performing.

When a task is rolled the task's impure segment, including user dedicated locations, and any task queue or message buffer structures are written to the roll file.

The memory occupied by the task's pure segment is not rolled and a resident task is naturally not rollable. According to a task's options, a nonresident task may or may not be rollable. Associated with every rolled task there is usually a roll file. To minimize I/O overhead, all roll files are normally contiguous and are transparent to the user. A roll file is deleted when a task goes end-of-task.

Each time memory is freed, the roll-queue is searched for the highest priority task whose memory requirements can be satisfied by the available memory or by the memory available and then freed by rolling a lower priority task. If such a task is found, it is rolled into memory. To control the roll-queue, the operator and the intertask communication call may modify the priority of rolled tasks. Arrangements are also made for a Supervisor Call or an operator command to CANCEL a rolled task.

(c) System Space

System Space is an area of local memory an operating system uses for tables, buffers, and data structures required for operation of the system. Data structures normally allocated to system space are task control blocks, file control blocks, access control blocks, and timer queue elements. An initial system space size is usually set at SYSGEN time and size requirements depend upon the total number of tasks in memory and the disc files to be open at any one time.

The use of system space must be controlled so a user-task cannot seize excessive space. At task establishment time, a limit is set for each task indicating the amount of system space it can request.

(d) Sample Memory Map

Fig. 5.8 represents a possible snap shot of the allocation of memory in a typical operating system. The configuration is a processor with 512KB of local memory and 64KB of shared memory. The OS is presumed to occupy 64KB.

5.4.5 Spooling

An operating system often provides a spooler task for input and output spooling. Operating system facilities to support the spooler task can be included as system generation options.

(a) Output Spooling

At system generation time, specific device mnemonics can be designated as pseudo-print devices. When the spooler task is started, the start options specify linkages between pseudo-print devices and physical-print devices. The physical-print devices become exclusively assigned to the spooler task and cannot be used by other tasks, except through the spooler.

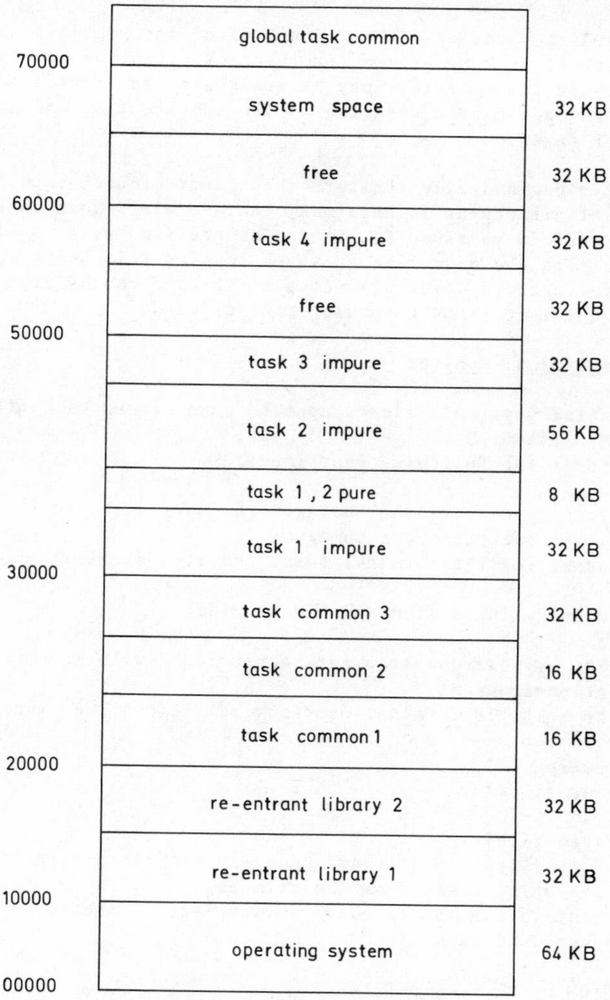

global task common	
system space	32 KB
free	32 KB
task 4 impure	32 KB
free	32 KB
task 3 impure	32 KB
task 2 impure	56 KB
task 1 , 2 pure	8 KB
task 1 impure	32 KB
task common 3	32 KB
task common 2	16 KB
task common 1	16 KB
re-entrant library 2	32 KB
re-entrant library 1	32 KB
operating system	64 KB

70000
60000
50000
30000
20000
10000
00000

Fig. 5.8 Example memory map

The operating system intercepts assign requests to pseudo-print devices and automatically allocates and assigns disc files to the specified logical unit. The operating system uses a special naming convention for spool files which reside on the spool volume. When the logical unit to which the spool file is assigned is closed, the operating system sends the file name, task name and priority of the task which requested spooling to the spooler task. The spooler task keeps this information on the spool print queue, a disc file assigned to and maintained by the spooler task. The print queue is searched by the spooler task for files to be printed. Files are printed in task priority order. If a file is found that needs to be printed, it is copied to the physical-print device.

(b) Input Spooling

Input spooling provides the inputting of information directly to a disc file without user programme intervention. When the spooler task is started, certain input devices may be designated as input spool devices. These are then exclusively assigned to the spooler task and may not be used by other tasks.

The spooler periodically monitors the status of each input device. If device unavailable status is returned, the spooler goes into a time wait. The read is reissued at the end of the time wait. A control card usually specifies the disc file to which the input is being spooled and when encountered the spooler allocates a file and copies from the device to the file until it finds an ending control card.

5.4.6 Multi-Terminal Monitor

Most operating systems also support some form of Multi-Terminal Monitor (MTM) system.
The MTM performs the following functions:

* it controls user directed I/O between the user task and the terminal
* it accepts and executes user commands
* it loads tasks for the terminal task, and it controls those tasks

Together the operating system and MTM provide:

* a real-time system environment concurrent with a multi-terminal development environment
* support for multiple terminal users using time-slice scheduling and Roll-in/Roll-out
* file protection
* private user files
* group files
* global system files
* the user facility for editing and creating files at the terminal and serially executing tasks from the terminal
* a concurrent batch background environment, capable of executing several background tasks.

5.5 CONCLUSION

In this chapter the main requirements of a computer system to support real-time process control have been considered. The majority of modern computers have the necessary hardware to support real-time operation but not all have adequate operating systems to enable real-time systems to be programmed in high level languages. There are also few systems available which permit the construction of simple operating systems programmed predominantly using high level languages.

5.6 REFERENCES

1. ALLWORTH, S.T.: 'Introduction to real-time software design', (MacMillan Press Ltd., 1981).
2. BARRON, D.W.: 'Computer operating systems', (Chapman and Hall, 1971).

3. BIBBERO, R.J.: 'Microprocessors in instruments and control', (Wiley, 1977).
4. CADZOW, J.A., MARTENS, H.R.: 'Discrete time and computer control systems', (Prentice Hall, 1970).
5. CULLEY, J.C.: 'Computer interfacing and online operation', (Edward Arnold, 1975).
6. CUTTLE, G., ROBINSON, P.B.: 'Executive programme and operating systems', (MacDonald-Elsevier Computer Monographs, 1970).
7. DORF, R.C.: 'Time domain analysis and design of control systems', (Addison-Wesley, N.Y., 1964).
8. ECKMAN, D.: 'Automatic process control', (Wiley, 1968).
9. FREEDMAN, A.L., LEES. R.A.: 'Real time computer systems', (Edward Arnold, 1977).
10. HEALEY, M., HEBDITCH, D.: 'Minicomputer in on-line systems', (Prentice Hall, 1981).
11. HOLT, R.C., GRAHAM. G.S., LASOWSKA, E.D., SCOTT, M.A.: 'Structured concurrent programming with operating systems applications', (Addison-Wesley, 1978).
12. HOPGOOD, F.R.A.: 'Compiling techniques', (MacDonald Computer Monographs, 1969).
13. KATZAN, H.: 'Advanced programming : Programming and operating systems', (Van Nostrand Reinhold, 1970).
14. KATZAN, H.: 'Operating systems : A pragmatic approach', (Van Nostrand Reinhold, 1973).
15. LEE, S.C.: 'Microcomputer design and applications', (Academic Press, 1977)
16. LIENTZ, B.P.: 'An introduction to distributed systems', (Addison-Wesley, 1980).
17. LISTER, A.M.: 'Fundamentals of operating systems', (Macmillan, 1975).
18. LONGBOTTOM, R.: 'Computer system reliability', (Wiley, 1980).
19. LOWE, E.I., HIDDEN, A.E.: 'Computer control in process industries', (Peter Perigrinus, 1971).
20. RAGAZZINI, J.R., FRANKLIN, G.F.: 'Sampled data control systems', (McGraw-Hill, N.Y., 1960).
21. SAGE, A.: 'Optimum systems design', (Prentice-Hall, N.J., 1968).
22. SAVAS, E.S.: 'Computer control of industrial processes', (McGraw-Hill, 1965).
23. TOU, J.T.: 'Digital and sampled data control systems', (McGraw-Hill, N.Y., 1960).
24. WEITZMAN, C.: 'Distributed micro-minicomputer systems', (Prentice-Hall, 1980).
25. WOOLVET, G.A.: 'Transducers in digital systems', (Peter Perigrinus, 1979).

Chapter 6

Communications for distributed control

6.1 INTRODUCTION

A Distributed Control System [1,2,3,] is one in which multiple
processors (microcomputers) cooperate to achieve an overall goal - the
control of a physical process or plant. The microcomputers may be
physically adjacent to the plant being controlled and exchange
information via a communication system in order to coordinate their
activities. They may be used to perform the following functions:

* Controlling sensors or actuators - at present, economics dictate that
  a number of sensors or actuators are connected to a microcomputer, but
  eventually single chip processors will be incorporated into the
  electronics of each sensor or actuator. For example the microcomputer,
  in an intelligent sensor, would do analogue to digital conversion,
  calibration and linearization of the sensor signal.
* Implementing control algorithms - microcomputers may be used to close
  the loop between sensors and actuators, replacing hardwired 3 term
  controllers etc., or they may perform more complex control algorithms
  e.g. involving optimization over a number of control loops.
* Operator interaction - microcomputers control visual display units,
  printers, keyboards, or special consoles. These are used to
  communicate with a human operator. The operator inputs setpoints,
  parameters or requests for data, and plant state information and
  alarms are output to the operator. This type of intelligent interface
  is rapidly replacing the array of dials and buttons found on the
  traditional control room console, as it is much simpler and easier to
  use.
* Communications - microcomputers can be used in traditional centralized
  computer control systems or remote monitoring systems for the
  transmission of information between the plant and computer. This
  allows the many wires to be replaced by a serial transmission system,
  which can also perform error detection, and pre-processing of the
  data.

6.2 ADVANTAGES OF DISTRIBUTED CONTROL

a) Costs

The use of serial communication lines rather than many parallel wires
from sensors and actuators to a centralized computer can result in
considerable cost saving in many applications e.g. in aircraft where

weight is important or in mines and large industrial sites, where distances between components can be thousands of metres.

The use of LSI circuits (microcomputers) to replace complex relay or hardwired logic can reduce both design and manufacturing costs.

The purchase price of multiple microcomputers will usually be much less than that of a powerful minicomputer which is traditionally used for centralized control systems.

b) Modularity

The use of independent processing modules, which communicate by sending messages enforces very clean interfaces. This reduces the complexity of the software which accounts for an increasing proportion of the implementation costs in modern systems. There is no need for complex operating systems within microcomputers. To a certain extent the additional complexities of communications offsets the simpler operating systems. An additional advantage of the modularity is the ease of expansion. New functions or capabilities can be included within the control system by the incremental addition of microcomputers without the cost of having to double the computing power or upgrading to a large computer.

c) Performance

The parallelism of multiple processors means a distributed control system can cope with much higher performance requirements than a centralized one. There is no need to scan large numbers of sensors using a single computer. Microcomputers can be dedicated to a particular task and so can easily cope with device or plant response times in the order of milliseconds [4]. In addition, microcomputers can be used to preprocess logged data and so reduce the amount of information transferred, and the computer's memory can act as a buffer to smooth bursts of traffic.

d) Reliability

It should be possible to design the system to continue working (possibly with reduced capability) after components have failed. The cost of including redundancy within the control system or having spare replacement computers is not particularly high.

The distributed intelligence can also be used to correct communication errors and so provide very reliable information transfer between actuators, sensors and controllers.

e) Development and Maintenance

The inclusion of intelligence within components of the control system makes it easier to develop parts of the system in isolation. A section of the plant can easily be isolated from the rest for testing or maintenance purpose. The control system can perform machine health monitoring e.g. detecting a gradual increase in a bearing temperature and give warnings of the need for preventive maintenance. The control system should include self-test and remote diagnostic capabilities.

The main problems in implementing distributed control systems are:

a) Communication System Complexity – the complexity and problems of implementing communication systems are often underestimated.

b) Lack of standards – there are very few accepted standards which would allow equipment produced by different manufacturers to be connected together in a single distributed control system. Some of the standards being developed are discussed at the end of this chapter.

6.3 NETWORK CONFIGURATIONS

6.3.1 Network Classification

This section compares the main configurations for interconnecting processing nodes or stations to form a network [5]. There are 2 classes of networks:

* Store and Forward – a network in which a complete packet or block of information is received into a buffer in the memory of an intermediate station before being retransmitted on the route to its destination. In general the stations are interconnected by independent point-to-point transmission lines.
* Broadcast – a network in which a message transmitted from one station is received by all other stations. In general this implies that all stations are connected to a common transmission medium and so a particular station must detect that a message is addressed to itself. All stations are directly connected by the common transmission medium and so communication does not involve an intermediate station.

6.3.2 Example Configurations

a) Complete Interconnection

Each station is connected by a dedicated point-to-point line to every other station as shown in Fig. 6.1.

Advantages:

* High integrity – failure of any one station should not affect communication between others; many alternative paths are available if a link fails.
* High throughput – multiple links can operate in parallel.
* Low delay – no intermediate stations i.e. no need for store and forward operations unless a link fails.
* Simpler communication as no routing problems.

Disadvantages:

* Very expensive – n stations require $(n - 1)*n/2$ full duplex lines and each station must have n–1 ports for all the lines.
* High expansion cost – adding the mth station requires m–1 additional lines and a spare port in all other stations.
* Broadcast messages are difficult to implement.

Fully interconnected networks are not very common because of the high cost. These are sometimes used for small numbers of stations.

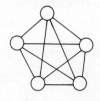

(a) complete interconnection

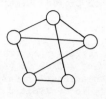

(b) mesh or partial interconnection

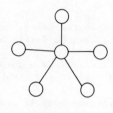

(c)   star

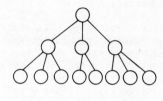

(d) tree or hierarchy

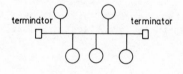

(e) highway or bus

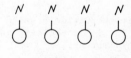

(f) radio network

(g) unidirectional loop

(h) bidirectional loop

Fig. 6.1 Network configurations

b) Mesh (Partial Interconnection)

A network with point-to-point links between stations, but all are not directly connected i.e. store and forward transmission is required between some pairs of stations. Each station should be connected to at least 2 others in order to provide alternate paths. This type of network is very common for large geographically dispersed networks as it is possible to provide high integrity at comparatively low cost.

Advantages:

* High integrity – alternate paths are easily provided in case of failure of a link or node.
* High throughput – parallelism of multiple links means the network is unlikely to be limited by link capacity.
* Very flexible – it is possible to match the placing of links and their capacity to the traffic requirements.
* Fairly low expansion cost – only 1 or 2 extra links are required for additional station.

Disadvantages:

* Delay – the network is store and forward and so the delay depends on the number of intermediate stations, this can be fairly high for a large network.
* Complex software – routing strategies can be complex and sophisticated: both point-to-point and end-to-end controls are required for efficiency and reliability
* Congestion – the store and forward nodes are susceptible to queuing delays and traffic congestion.
* Broadcast communication not easily implemented, as it requires sending a copy of message to every station.

c) Star

All communication between remote stations is via a central switching node. It is sometimes used to link remote intelligent stations to a central computer.

Advantages:

* Cheap – a single link is required to each remote station, which also results in a low expansion cost.
* Simple – very simple table look-up routing in the central switching node.
* Low delays – a maximum of 1 intermediate node.

Disadvantages:

* Poor integrity – a failure of a link isolates a station and the failure of the central switch stops all communication, and so redundancy is sometimes provided at the central switch. Note, failure of a remote station does not affect the rest of the network.
* Throughput is limited by that of the central switch, which may be a bottleneck.
* The network is usually not homogenous in that the central switch is different from the remote stations i.e. all stations are not interchangeable.

d) Tree or Hierarchical Network

This is really an extension of the star topology and so has very similar characteristics. It is often used in control systems, in spite of reliability problems, as it reflects the hierarchical nature of many control systems.

e) Highway

Also known as a bus or multidrop link. It is a broadcast system in which the shared transmission medium (which could be twisted pair, coaxial cable, optic fibre etc) interconnects all the stations. Only one station can transmit at a time and so the system must incorporate some technique for preventing or resolving contention for the transmission medium. Contention occurs if two or more stations attempt to transmit at the same time.

Advantages:

* Low wiring costs — a single transmission line links all stations which minimises the length of wire.
* Low expansion cost — fairly simple to tap-off signals at any point along the transmission medium.
* Easy reconfiguration — adding or removing stations is very easy.
* Simple software — no routing as communication is broadcast. Also error control is end-to-end, simplifying protocols.
* The transmission medium can be completely passive and so be inherently reliable.
* Each station can have equal status, although in some systems one station is made the master.

Disavantages:

* Integrity — all communication fails if the transmission medium is cut, and so redundant transmission lines may be required.
* High capacity transmission lines are required to cater for the sum of all the communication within the network.
* Contention avoidance techniques can be rather complex.
* Half duplex transmission — it is not usually possible to have full duplex transmission with a single line.
* The IEC (International Electrotechnical Commissions) favour the use of serial highways for process control as they are passive.

f) Radio Network

This is conceptually identical to a highway but uses radio waves rather than wires or cables e.g. Satellite Broadcast systems. Most of the characteristics of highway apply to radio networks.

Advantages:

* No wiring costs
* Transmission medium cannot be cut

Disadvantages:

* Expensive receivers and transmitters

g) Unidirectional Loop

Each station is linked to its neighbour by a simplex link and so communication is only in 1 direction round the loop. The loop interface usually regenerates the signal and contains a few bits of buffering (1 to 16 bits) and so a loop is known as a check and forward system. Transmission could be broadcast in that a message will go all round the loop and be removed by the sender or it might be removed by the destination. Loops are becoming increasingly popular for local networks.

Advantages:

* Low wiring costs.
* Low expansion cost - only 1 additional link is required for an additional station.
* Simplicity - no routing, and error control can be end-to-end.
* Broadcast transmission easily implemented.
* Delays are small provided the buffer in the loop is small.

Disadvantages:

* Poor integrity - the failure of any link or station could stop all communication unless redundancy is included. The loop interface should be designed so that if a station fails (e.g. power failure), the interface bypasses the station which is then isolated from the loop.
* Some form of contention resolving mechanism is needed.
* The signal is usually regenerated at each loop interface, which means the transmission line is not passive, i.e. it consists of a number of point-to-point circuits.

h) Bidirectional Loop

A full duplex link connects each station to its neighbour. This type of loop can either operate as 2 unidirectional loops - an ordinary unidirectional loop plus a redundant one in case of a failure or as a store and forward mesh network with a connectivity of 2.

6.3.3 A Comparison of Store and Forward and Broadcast Configurations

There is considerable controversy in the process control world as to which type of network configuration is best. The IEC have stated in their requirements for a communiation system for process control [10], that it must not be store and forward, without justifying this decision. The following table outlines the arguments for the 2 classes of networks.

Broadcast	Store and Forward
(Highways, radio networks, some loops)	(Star, Tree, mesh, some loops)
No routing hence simpler software.	Routing Techniques can be very complex.
All error control is end-to-end.	Point-to-point controls are required for efficiency and end-to-end controls are needed to cater for intermediate node failures.
The destination address must be recognised before a message can be received by a station. So the address must be one of the first fields in the header.	The information coming in on a point-to-point line is first received then the destination address can be checked at the message has reached its destination or should be routed on.
A single transmission medium must support the sum of all the communication within the network, thus high speed lines (usually 1-10 Mbits/sec) are required.	Multiple transmission lines can operate in parallel and so lower speed lines (2K - 50K bits/sec) can be used.
Line interfaces may have to be constructed out of MSI or SSI logic and may have to include buffers if the data rate is too high for DMA access.	Cheap LSI interfaces which provide some error control facilities are available
The main component of the delay is waiting for access to the shared transmission medium (contention delay).	The main delays are due to transmission times over multiple hops.
The transmission path can be completely passive and so be inherently reliable.	Transmission path includes switching nodes so is less reliable.
Redundant transmission lines and duplicate expensive line interfaces may be required in case a line is cut.	Alternate paths are easily achieved in mesh networks.
Low wiring costs, A simple system can be built with a single interface per station.	More wiring required and mesh networks require at least 2 interfaces per station.
May not be compatible with PTT lines.	Compatible with PTT lines.
Suitable for local networks.	Suitable for geographically dispersed large scale networks.

In general higher hardware costs
for interfaces and high speed
lines.

In general more complex communication software.

6.3.4 Interconnected Networks

The previous section has shown the problems of deciding that one
particular network configuration will suit all applications. There is a
need to support both store and forward and broadcast communication,
because of the hierarchical nature of a typical control system on a
large site.

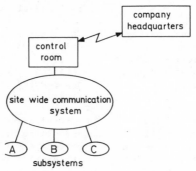

Fig. 6.2 Hierarchical structure of a control computer

It consists of a number of subsystems (Fig. 6.2), which must be
interconnected by a site wide communication system to allow overall
plant scheduling or optimisation. A subsystem may manufacture a
component e.g. it might be a boiler in a chemical plant. The subsystem
could consist of a number of complex machines with their own control
systems. This structure is best represented by a number of independent
subnetworks which are interconnected by a site-wide communication
system. There may be additional point-to-point lines required to meet
particularly stringent performance requirements e.g. to close a control
loop between 2 subsystems. A possible interconnected network
configuration is shown in Fig. 6.3. The choice between a loop or highway
for the subnetwork will depend on the LSI circuits which are emerging
for the applications such as office automation.

6.3.5 Station Structure

A station or node within a network is itself likely to consist of a
cluster of microcomputers. These are physically close (within the same
cabinet) and are closely-coupled, communicating via shared memory or a
parallel bus (see Fig. 6.4). A discussion of closely coupled
interconnection structures will not be given in this chapter but can be
found elsewhere [5].

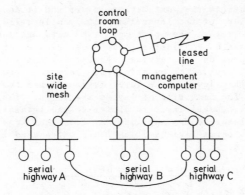

Fig. 6.3 A typical interconnected network structure

The advantage of including more than 1 process in a station is that it allows the processing power to be matched to the application requirements. A control loop may require information from a large number of sensors and may involve complex calculations. In addition the serial communication function can be physically separated from the control application processing.

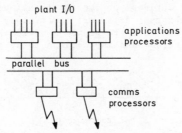

Fig. 6.4 A possible station structure

This structure results in modularity at a number of levels:

* An overall system consists of subnetworks
* A subnetwork contains stations which can be added as required
* A station's processing capability can be enhanced by adding additional processors or input/output
* A processor itself may contain more than one task (process).

6.4 LAYERED NETWORK STRUCTURE

6.4.1 Components of a Layered Structure

A communication network can be extremely complex and so it is organized in a 'hierarchical' or layered structure [6,7] similar to that found in many operating systems.

a) Layer

A layer consists of a set of service units (entities) which

provides services to the next higher layer, and it does so by making use of the services of the layer immediately below. The service units constituting a layer are part of the data stations forming the network and so can be considered to be a distributed subsystem which cooperate to provide a service. The service units forming a layer communicate with each other by using the services provided by the lower layer. The _level_ of a layer is an ordinal number associated with the layer. The numbering of layers startS at the transmission line and so a 'lower' layer is closer to the transmission line and higher one is closer to the application, i.e. the transmission line has level 0. The terms layer and level are often used interchangeably in the literature.

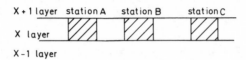

Fig.6.5 A layer

b) Protocol

A protocol is the set of rules governing communication between the functional units which constitute a particular layer (Fig. 6.6). That is the X protocol defines how the functional unit at the X level in one station exchanges data or supervisory information with the X level at another station. The protocol defines the format of the information exchanged and any actions to be performed on receipt of the information.

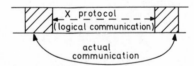

Fig.6.6 A protocol

c) Interface

The interface between two layers defines the means by which one layer makes use of the services provided by the lower layer (Fig. 6.7). It defines the rules and formats for exchanging information across the boundary between adjacent layers within a single station. The interface may be specified in terms of its mechanical, electrical, timing or software characteristics i.e. the interface may be physical or logical.

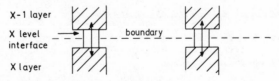

Fig.6.7 An interface

d) Information Units

   The functional units forming a layer communicate with each other by exchanging units of information which may be fixed or variable length. These are called protocol data units and correspond to messages, packets, frames etc. We use the term frame as a general term to refer to a protocol data unit at any level.

   In order to transfer an X+1 level frame according to the X+1 level protocol, the X+1 layer must pass the frame across the X interface to the X level. The unit of information passed across the interface is called an X interface data unit and in general does not correspond exactly to the X+1 level protocol data unit. The interface data unit may contain additional control information indicating the type of service required e.g. the address or name of the destination, or a timeout period for which the X+1 layer will wait for a reply, an indication that no reply is expected or a priority level of the information.

   The X layer sees the X+1 layer frame as a unit of user information which must be transported. In general it will add an envelope of its own control information consisting of a header and possibly a trailer. This then forms the X level protocol data unit or frame. The X+1 level frame should be completely transparent to the X layer, which may process it in some way to form an X level frame. For instance, user information unit from a higher level may be encrypted, bit-stuffed or fragmented into smaller units for transport. This concept of nested information units is shown in Fig. 6.8.

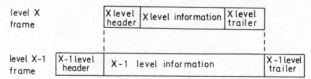

Fig. 6.8 Nested practical data units

6.4.2 Reasons for Layered Structure

   Viewed from above, a particular layer and the ones below it may be considered to be a 'black box' which implements a set of functions in order to provide a service (Fig. 6.9). This structure exhibits the following advantages associated with the concept of modularity or information hiding.

* Independence between layers

   A layer makes no assumptions about how the one below it is implemented, but only has knowledge of the services provided through the interface.

* Flexibility

   A change at any one layer, for instance due to a change in technology, should not affect the layers above or below it. A particular network may choose to implement only a subset of the layers and it can still be compatible with other networks which provide all layers.

* Physical Separation

The layers can be implemented using the most appropriate technology e.g. software package in an application processor, microprocessor software or LSI hardware. It has proved advantageous in most networks to implement the layers within a station using a minimum of 2 or 3 processors.

* Simplicity of Implementation

This structure makes the implementation of a complex communication system manageable because it decomposes the overall services into easily comprehensible sections.

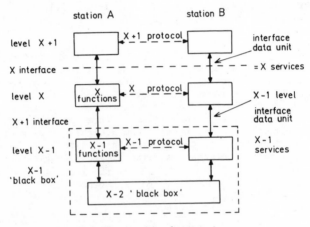

Fig. 6.9 Hierarchical structure

6.4.3 A Layered Network Architecture for Process Control

As discussed in section 6.3, there is no single network interconnection structure which is suitable for all process control applications. The network architecture should cater for both broadcast (serial highways or loops) and store forward networks (mesh or arbitrary point-to-point links). The main objective of the proposed network architecture is to allow a system designer to configure the network to suit the application using whatever interconnection structures are appropriate. It is still necessary to have standard protocols and interfaces within this network architecture to allow equipment from different manufacturers to be interconnected.

The network architecture proposed in this paper is based on the ISO Open System Architecture Reference Model [8]. It is being developed at Imperial College as part of a research project funded by the National Coal Board, to design a distributed computer architecture suitable for use in local mines. The architecture is an implementation rather than reference model and so the ISO model has been simplified by omitting the session and presentation layers. The functions performed by these layers are not required in a process control local network.

The proposed network architecture consists of the following layers:

* Application layer - this includes all tasks (processes) performing application dependent function e.g. controlling devices.
* Transport layer - this provides inter-task communication and isolates the task from the physical implementation of the network i.e. local and remote communication appears identical to the task.
* Network layer - this performs a routing function.
* Data link layer - this abstracts the characteristics of the different interconnection structures (loop, serial highway, point-to-point lines etc.) into single logical highway view.
* Physical layer - provides for the transmission of bit streams and hides the characteristics of the transmission media.

a) Application Layer

The application layer is programmed in high level language and contains modules and tasks which perform process control functions.

A task is a sequential programme, commonly called a 'process'. The word task is used to avoid confusion with the term 'process control'. A task could possibly handle a particular device e.g. analogue to digital converter.

A module is a collection of cooperating tasks in a single station probably related to a particular function e.g. controlling a pump. The module incorporates concurrency (i.e. multiple tasks) and is the largest unit of abstraction seen by a programmer. A station may contain more than one module.

Modules contain entryports and exitports which are used by tasks for communication, both within a module and between modules. A task receives messages on an entryport and sends messages to an exitport. An exitport is linked to one or more entryports to form a unidirectional association.

This may be performed statically during system installation or dynamically at run time. Ports are strongly typed i.e. only messages corresponding to the port type can be sent or received on a port. Priority is associated with ports rather than messages i.e. an alarm message would be sent on a high priority alarm exitport.

There is no shared data between tasks or any form of communication other than message passing. The ports provide a very clean visible interface to a module, similar to the I/O sockets on a hardware module. A more detailed description of these concepts is available in [9].

b) Transport Layer

The transport layer implements the intertask communication and so it performs some functions traditionally performed by an operating system or by a runtime system of a high level language which provides communication primitives. The transport layer isolates the application layer from how communication is implemented, whether over serial links, parallel bus or even shared memory. An application task is thus isolated

from any knowledge of network configuration or the location of a task
with which it communicates.

Functions Performed

* End-to-End Error and Flow Control – this is an end-processor to
  end-processor function. It is necessary because the communication path
  between 2 tasks may be via a number of intermediate communication
  links and communication processors.
* Name to Address Translation – the application task uses a logical name
  (module, port) which must be translated into a physical address so
  that a message can be delivered to its destination.
* Routing – a simplified form of routing is performed in this layer. If
  a message is destined for a module in the same processor, it is queued
  on the relevant entryport, i.e. all messages go into the transport
  layer but local messages come straight out again. The message may be
  to a module in a different processor, within the same station and so
  the transport layer passes it on via a shared memory or parallel bus.
  If the destination is a remote station then the message is passed to
  the communication processor.
* Priority is implemented by ordering of internal queues if necessary.
* The transport layer does not perform fragmentation or reassembly of
  messages i.e. an application message corresponds to a protocol data
  unit transferred over a communication link.

c) Network Layer

The network layer is responsible for getting a message from a source
station to a destination station. It is a comparatively simple layer as
there is no peer-to-peer protocol. The network layer isolates the
transport layer from knowledge of the network configuration.

Functions Performed

* Routing – this is the main function of the network layer. If the
  subnets are intrinsically broadcast (loops or serial highways) then
  routing is only needed between subnets. Because of the static nature
  of the application, fixed routing with alternative routes is adequate.
* Error Messages – if a message cannot be delivered at the destination
  because of node or link failure, the network level generates an error
  message to the original source of the message.

d) Data Link Layer

This layer transforms an error prone physical connection into an error
free one. It isolates the network layer from knowledge of the
interconnection technique used – whether a loop, serial highway or point
to point line. The network layer sees all these physical
interconnections as highways, each with one or more other stations on
it.

Functions Performed

* Error Control – this is accomplished by means of cyclic redundancy
  checks (CRC) and retransmissions.
* Access Control – a shared transmission medium e.g. loop, serial

highway, half duplex point circuit requires a mechanism for controlling access by each station wishing to use the medium. This architecture allows many different access control mechanisms e.g. polling, baton passing, or contention-based to be used.

* Address recognition – each station on a logical highway has a unique address, but the station may recognise more than one address (e.g. a broadcast address as well).

* Character synchronization – the layer also detects character boundaries as it is assumed that messages will consist of multiple 8 bit characters or bytes.

* Frame synchronization – detection of the start and end of a frame is performed at this level either by detecting special start of frame characters; or by detecting violations of the normal signalling method.

e) Physical Layer

This layer is responsible for all transmission medium dependent operations involved with transmitting a bit stream across a transmission line. It also performs some functions which logically are part of the data link layer but are best performed by the transmission hardware.

Functions Performed

* Modulation/Demodulation – this may involve the conversion between digital information and analogue signals or may simply involve level conversion for base band signalling.

* Bit synchronization – this level is responsible for detecting start and end of bits.

* Carrier sense – some data link access methods require detection of the carrier i.e. whether the data link is busy or not.

* Connection/Disconnection – this allows a station to be physically disconnected from the transmission medium e.g. in case of power failure or a station fault.

* Signal Quality Monitoring – this is an additional error detection service performed for the data link layer as the redundant coding (e.g. CRC checks) may not provide an adequate residual error rate for control applications.

6.5 NETWORK MANAGEMENT

A local network should incorporate facilities for the management of the communication system itself. This includes monitoring the performance as well as maintenance and diagnostic facilities. Part of the network management is distributed – each station contains a Network Control Centre (NCC) which is probably a part of the plant control room. Centralising the NCC does have reliability implications, but it is only responsible for long term functions and so its failure will not affect the networks capability for transferring messages.

6.5.1 Station Management Function

The network management module is very similar to an application task but it has special interfaces to the layers within the communication system to perform some of its functions.

* Monitoring – the management function collects information on the performance of the communication system, which it periodically sends to the NCC. Typical information maintained includes:

    Status of transmission lines
    Average lengths
    Number of message retransmissions
    Number of messages received in error
    Totals of messages received and transmitted

* Maintenance of tables – new routing or updates would be sent out by the NCC to those stations involved in routing.
* Diagnostics – the station includes facilities for loopback tests, dumping specified sections of memory etc.
* Bootstrap – this is the loading of code via the communication system. Although an operational system is likely to contain code in ROM, the communication system should still include a remote bootstrap facility for development and testing purposes.
* Error Reporting – each station should report errors such as communication line failures to the NCC. Of course a station isolated by line failures cannot report this to the NCC, but other stations would report the inability to communicate with the isolated station.

6.5.2 Network Control Centre Functions

* Collecting Statistics – the NCC collects and analyses statistics reports to indicate where maintenance is required.
* Generating of Routing Tables – the NCC maintains a picture of the whole network topology and distributes new tables or updates when the network is modified or when stations fail.
* Remote Diagnostics – the NCC provides facilities for an operator to perform remote diagnostics on a station.
* Coordinate Error Reports – received error reports are analysed and the operator informed of the action required e.g. if two stations both report failures of each other, this means the communication link between the two has failed.

6.6 STANDARDS ACTIVITY

There are no existing standards for communication systems for distributed process control, but a number of standards organisations are currently working on proposals which are directly applicable or relevant to control.

a) **International Electrotechnical Commissions (IEC)**

There are 2 IEC committees working on standards for communication systems.

i) TC65A, Working Group 6, has a draft proposal on the functional requirements for a 'Process Data Highway (PROWAY) for Distributed Process Control' [10]. This, so called, functional requirement is actually an outline design specification mixed up with the functional requirements. The design is for a serial highway with a single transferable mastership. Only one station can be master of the transmission medium at any instant, but the mastership can be transferred between stations. Transmission rates of 1 M bits/sec over a 200 m highway or 30 k bits/sec over 2000 m are mentioned, and the highway can have up to 100 stations. The document does not actually specify how stations access the shared transmission medium, but polling is implied. This committee has also specified a network architecture consisting of:

* Application Layer – similar to that described in the previous section.
* Network Layer – a mixture of the transport and network layers defined previously. This layer is rather vague, as it is really not part of the specification of the highway which does not include any store and forward transfers.
* Highway Layer – roughly corresponds to the data link layer i.e. error control and control of access to the transmission medium.
* Path Unit – this layer performs functions of frame and character synchronization, error code generation and checking.
* Live Coupler – essentially the same as the physical layer.

It is likely that PROWAY will be based on a modified form of the ISO HDLC protocol. The standard ISO version is inadequate because it will not give the required residual error rule; it does not include both destination and source addresses in the header and there is no mechanism for transferring mastership.

The PROWAY document also includes an extensive glossary, but the document has a number of inconsistancies. In my opinion any standards emerging from this source are unlikely to be very successful. A number of manufacturers have announced products which they consider are contenders for standardization e.g. Brown Boveri's Portner Bus, Foxboro's Foxnet.

ii) TC 57 is working on a document highlighting the requirements and conditions for data transmission systems for telecontrol. They are considering longer distances and lower speed (< 2000 bits/sec) than the other IEC committee. They also define a layered structure (different from PROWAY) but roughly similar to that described in this chapter.

b) <u>International Standards Organisation (ISO)</u>

TC 97/SC 16 has produced an open systems architecture [8, 11]. This is a reference model of a network architecture which should cover all forms of networks. This reference model will form the basis from which standard protocols are defined. A number of large companies (e.g. IBM, ICL, Bell, Bell Northern) as well as researchers and academics are represented on this committee.

They are working very rapidly for a standards committee. Their latest document is rather complicated to understand although the terminology used and concepts are completely consistent. The work of this committee is essentially aimed at interconnecting large independent computers for remote access, file transfers etc.

The network architecture consists of 7 layers – roughly the 5 presented in section 4.3 and 2 additional layers between the application and transport layer. These are a session and presentation layer, but they are not really applicable to process control.

c) Institute of Electronic and Electrical Engineers (IEEE)

They have established a local network standardising committee. Their application areas are likely to be office automation or resource sharing and they may go for an ETHERNET-like serial highway [12].

6.7 REFERENCES

1. FARBER, G.: 'Principles and Applications of Decentralised Process Control Computer Systems', IFAC, 1978, pp.385-392.
2. SYRBE, M.: 'Basic Principles of Advanced Process Control Systems and a Realisation with Distributed Micrcomputers', IFAC, 1978, pp.393-401.
3. IEE 'International Conference on Distributed Computer Control Systems', IEE Conference Publ. No.153, 1977.
4. PRINCE, S., SLOMAN, M.: 'The Communication Requirements of a Distributed Computer Control System', Imperial College Research Report No. CCD79/33, 1980.
5. WEITZMAN, C.: 'Distributed Micro/Minicomputer Systems', (Prentice Hall, 1980).
6. POUZIN, L., ZIMMERMAN, H.: 'A Tutorial on Protocols', Proc IEEE 1978, v.66, No.11, pp.1346-1370.
7. DAVIES, D., BARBER, D., PRICE, W., SOLOMONIDES, C.: 'Computer Networks and their Protocols', (Wiley, 1980).
8. ISO/TC97/SC16N227: 'Reference Model of Open Systems Interconnection (Version 4)', 1979.
9. LISTER, A., MAGEE, J., SLOMAN, M., KRAMER, J.: 'Distributed Process Control Systems: Programming and Configuration', Imperial College Research Report No.80/12, 1980.
10. IEC/65A (SECRETARIAT) 18/WGG. 'Process Data Highway (PROWAY) for Distributed Process Control Systems: Part 2 Functional Requirements', Report No.79/23908, 1979.
11. ZIMMERMAN, H.: 'The ISO Open Systems Interconnection Architecture', Real Time Data Handling and Process Control, (North Holland,1980), pp.433-442.
12. METCALFE, R., BOGGS, D.: 'Ethernet: Distributed Packet Switching for Local Computer Networks', CACM 1976, v.19, No.17, pp.395-404.

Chapter 7

Languages for computer control

7.1 INTRODUCTION

This chapter concerns the requirements for and availability of programming languages for embedded-computer systems — that is, for systems in which a computer or computers, (mini, micro or mainframe) are directly connected for real-time use to control or instrument some other equipment.

Real-time programming is substantially more difficult than conventional programming. This chapter considers the reasons why that is so, and what assistance the programming language can provide to the programmer in coping with these difficulties. It also looks ahead slightly to proposed schemes for programming support environments, in which comprehensive software tools are brought together to help the programmer.

The fundamental point is that programming for interaction with the real world is both more complicated and less forgiving than programming for conventional numerical calculations. We can identify three areas of difference which must be covered in real-time programming:

* multiple concurrent activities
* unconventional input/output
* programmed contingency actions.

As well as these directly technical issues (which will be addressed in more detail presently), there are further requirements which arise because of the complexity and size of real-time programmes:

* team-built programmes
* well-validated programmes

(in other words, Software Engineering requirements).

The most comprehensive and up-to-date statement of requirements for programming languages for embedded computer systems is the Steelman document [1]. Earlier work on the development of programming languages took place without explicit identification of the requirements, and consequent loss of clarity. Two real-time languages (Coral 66 and RTL/2), will be mentioned briefly and more details about another (Modula) will be given. ADA is a future language which will be important in the latter half of this decade.

7.2 CORAL 66 AND RTL/2

Coral 66 [2] was invented as a derivative of Algol 60, with some influence from Fortran for team programming. It barely addressed the main problems, because they were not understood at the time. Its distinctive features are its ability to handle data representations: bit manipulation and absolute addresses. These are needed to support unconventional input/output, although the input/output operations are not themselves expressable. It has no multi-programming or exception handling facilities. Its other differences from Algol 60 concern performance; a permanent worry among real-time programmers is to get the programme fast enough, and there is a natural reluctance to use features which carry a run-time overhead. Algol 60 storage allocation and procedure calls require run-time action, so the arrangements in Coral 66 were changed to make these static (and more like Fortran).

RTL/2 [3] was a serious response to the technical requirements. It provides for multi-programming and (in a limited way) for unconventional input/ouput. It has no special facilities for contingency programming. The structure of programmes is suitable for team work, but no particular help is given for programme validation. RTL/2 is broadly based on Algol 68, and accordingly has programmer-definable modes (or types as we now call them) which greatly improve the legibility (and hence correctness) of a programme. Its textual structure consists of a number of units called bricks – procedures (re-entrable), data (shareable) or stack (one per task). Each task runs concurrently with the others, and has its own stack brick. A task has a root procedure which may call other procedures (not necessarily distinct between tasks) and refer to data bricks (which may be private to the task or shared between several tasks). The number of tasks is static, and the overhead on task-switching is slight. Storage allocation may be static (in data bricks) or dynamic (on the stack), depending on the requirements of the programme.

7.3 MODULA

Modula was invented by N.Wirth [4] to provide a programming language for what was hitherto held as the bastion of assembly language programming – small computers in control systems. Modula provides a very powerful and neat multi-programming facility, with concern for programme validation underlying the design. The main technical parent is Pascal, but many features of Pascal were dropped (to make the compiler simple), and major new ideas were incorporated.

The principle new idea in Modula has nothing at all to do with real-time, but is a basic Software Engineering idea: programme modularity. A module is a collection of declarations (constant, type, variable, procedure) with explicit programmer control of access to and from the rest of the programme. Note how this contrasts with the (traditional) block structure of Algol 60, Algol 68 and Pascal: in those, there is no explicit control of access, and each declaration is visible throughout its containing block with no protection. Modula uses this technique to achieve a simple structure in quite complicated programmes.

Example 1

This example is from Wirth [4]. Assuming that an object, a, is declared in the environment of M1, then the following code also makes b and c accessible in this environment:

```
MODULE M1;
 DEFINE b,c;
 USE a;
 {DECLARE d}

 MODULE M2;
  DEFINE c,e;
  USE d;
  {DECLARE c,e,f}
  {c,d,e,f are accessible here}
 END M2;

 PROCEDURE b;
  {DECLARE f}
  {a,b,c,d,e,f are accessible here}
 END b;

 {a,b,c,d,e are accessible here}
 END M1
```

The identifiers in the define-list are 'exported', those in the use-list are 'imported'. It should be noted that there are restrictions on the use of objects which are exported e.g. variables can only be changed in the module in which they are local, outside this module they are read-only variables.

Another idea in Modula is that a process is very much like a procedure. Indeed, they are declared in just the same way (apart from the introductory keyword), and are also called in the same way. The difference is in the semantics. When a procedure is called, the caller waits at the procedure call statement until the procedure body has been executed, and then continues executing from the call statement. When a process is called, the caller sets the corresponding process body in execution but <u>does not wait</u> until that reaches completion (which it need never do) - the caller continues executing from the call statement in parallel with the called task.

Previous designers of multi-programming systems have pointed out dangers of programming errors among tasks, and Modula was cunningly designed to avoid these. An important requirement for safe programming is to be able to have a serially reusable procedure - one which can be entered from several tasks but not at the same time, and not entered from one task while it is being executed for another. (These are called monitors in the literature). Modula combines the ideas of module and procedure to achieve this effect. A special kind of module, called an interface module, may contain a number of procedures (but not processes). Between such points, it is guaranteed that no more than one of the procedures in the module will be in execution at any time. This provides mutual exclusion.

Synchronisation between tasks is provided by special data types called signals. These are simpler than the classical semaphores of Dijkstra [5] in that they imply no memory and can be implemented with very little overhead (a few machine instructions). Any data object of type 'signal' may be used only in three kinds of operation: wait (which holds up the current process); send (which allows at most one currently waiting process referring to that object to resume); and awaited (which indicates whether or not there are any processes currently waiting for that signal to be sent). Signals are used for all synchronising purposes – principally to control the usage of buffers between tasks, to ensure that a consumer task does not get ahead of a producer task.

Another major advance in Modula is its scheme for dealing with unconventional input/output. This is of course machine specific, and Modula as defined deals with the PDP-11. The basic idea is that the interfaces between the computer's central processing unit and the device controllers should be accessible to the programmer. These interfaces are particular registers at particular Unibus addresses, e.g. variables may be declared as follows:

```
VAR    kbs[177560B]:bits    {keyboard status}
       kbb[177562B]:char    {keyboard buffer}
```

where [177560B] and [177562B] are explicit addresses.

Recognising that device programming is particularly tricky and prone to error, Modula insists that it should be confined to identified modules, called device modules. These have additional facilities allowed: they have an explicit priority (for the CPU priority at which the device handlers are to run), and may include data declarations which give the absolute Unibus address of their object. Within a device module, there may be processes (which are the device handlers), and these may contain the statement 'doio' (which means suspend the CPU process until the I/O device has completed its action and indicated that fact by sending an interrupt). The device process has the interrupt vector address in its heading. Apart from these extensions, device modules are like interface modules.

Example 2

A device with priority 5 and with a device process with interrupt response vector at location 364B.

```
DEVICE MODULE    scope [5];
DEFINE    pulse, reset_pulse, tick;
VAR       pulse : boolean;
          tick : signal;
          event_link [170442B] : bits;
CONST     may_interrupt = 14;
          busy = 6;

PROCEDURE reset_pulse;
BEGIN     pulse := false
END       reset_pulse;

PROCESS synchronise [364B];
```

```
BEGIN     event_link [may_interrupt] :=true;
          LOOP
              event_link [busy] := false;
              doio;
              pulse := true;
              IF awaited (tick) THEN
                  send (tick)
              END
          END
END       synchronise;
BEGIN     {scope}
          synchronise
END       scope;
```

For an assessment of Modula see Holding and Wand [6].

7.4 ADA

During the last five years there has been world-wide effort, stimulated by the US Department of Defense, to produce a language which could become a common language for embedded computer systems. This has now been defined, and is called ADA [7]. The aim of the whole project was (and is) to reduce the high cost of software – estimated in 1975 to be $3x10^9$ dollars per annum. Many factors contribute to the cost which a common language would overcome: re-inventing wheels for each project, shortage of software tools, and inability to re-use already written software were the principal sources of waste in new software, but the major cost overall is on software maintenance: up to 80% of the lifetime cost occurs after the software has gone into service, and has to be modified (because of operational changes) by people other than the original author. Hence a primary goal of ADA was to put readability before writability: it is more economical to have the programme written with clear statements of design assumptions, even though this implies more time and effort at the programme writing stage. A piece of programme is written only once – it is read many times.

The current state of ADA is that the initial language [8] has been subjected to extensive test and evaluation, as a result of which the revised version has been produced. The official 'ADA Debut' was on 4th and 5th September 1980. A number of implementation projects are now under way, to produce compilers for the language. Partial compilers are expected in 1981, full compilers for experimental use in 1982, and serious production compilers in 1983/84. Projects which will reach the programming stage from 1985 onwards are expected to use ADA.

It is important to realise that the US High Order Language project is not only about the ADA Language, but about the Programming Support Environments which will encompass it. This will be a comprehensive set of compatible software tools, together with a project database and suitable man-machine interfaces, to allow for the production, maintenance and validation of ADA programmes and management of software projects involving ADA. Work on the ADA Programming Support Environment (APSE) is at an earlier stage than for ADA itself – a statement of requirements has been published [9] and a contract to design an APSE has been awarded. The APSE is planned to be open-ended, so that a relatively modest set of tools can be provided in the early versions, to be enhanced with the passage of time. APSE will itself be written in ADA, so that it can be mounted on any (sufficiently powerful) computer having an ADA compiler. Current estimates are that an ADA compiler will need more than a 16-bit address space, so will not be feasible on a PDP-11 or current generation of microprocessor, but could work on the next generation of 32-bit micros and minis. The major requirement then will be for adequate backing store to hold the project database.

7.5 REFERENCES

1. 'Steelman': Programming Language Requirements for Embedded Computer Systems, US Department of Defense, 1979; See also PYLE, I.C., 'Methods for the Design of Control Software', Software for Computer Control Proceedings Second IFAC/IFIP Symposium on Software for Computer Control Prague 1979, (Pergamon,Oxford,1979) pp.51-57.
2. WOODWARD, P.M., WETHERALL, P.R., GORMAN, B.: Official Definition of CORAL 66, (HMSO,London,1970).
3. BARNES, J.G.P.: 'Real-time Languages for Process Control', Computer Journal, 1972, v.15, pp.15-17; BARNES, J.G.P.: RTL/2: design and philosophy, (Heyden,London,1976).
4. WIRTH,N.: 'Modula: a Language for Modula Multiprogramming', Software - Practice and Experience, 1977, v.7, pp.3-35; WIRTH, N.: 'The use of Modula', Software - Practice and Experience, 1977, v.7, pp.37-65; WIRTH, N.: 'Design and Implementation of Modula', Software - Practice and Experience, 1977, v.7, pp.67-84.
5. DIJKSTRA, E.W.: 'Cooperative Sequential Processes', in Programming Languages, (F.Gennys,Editor), (New York,Academic Press,1968).
6. HOLDING, J., WAND, I.C.: 'An assessment of Modula', Software - Practice and Experience, 1980, v.10, pp.593-621.
7. Reference Manual for the ADA programming language, US Department of Defense, 1980.
8. ICHBIAH, J.D. et. al.: 'The ADA Programming language', SIGNPLAN Notices, June 1979.
9. STONEMAN (revised): Programming support environment requirements, US Department of Defense, 1979.

Chapter 8

On-line computer control of pH in an industrial process

8.1 INTRODUCTION

Feedback control of acidity/alkalinity is notoriously difficult. The difficulty arises because ionic concentration, which determines acidity/alkalinity, cannot be directly measured. Usually it is only practicable to measure pH (puissance d'Hydrogen), [1] which takes the form of a potential difference in an electrolytic cell and is related to the concentration $[H^+]$ of hydrogen ions (in kg mol m^{-3}) by

$$pH \equiv -\log_{10} [H^+] \qquad (8.1)$$

It is the severe, logarithmic nonlinearity of eqn. (8.1) that makes feedback control difficult.

Biological processes are very sensitive to acidity/alkalinity, so there are many applications where feedback control of pH is required. These include maintenance of correct conditions in production processes and neutralisation of industrial effluent so that it shall not harm the environment. Limits on the allowable pH of the effluents are imposed by legislation [2] and are being enforced with increasing stringency.

This chapter reports on research into the contribution which modern control technology, in the form of an online digital computer using recently-developed control algorithms, can make to pH control. The research is part of a continuing collaborative project, between the Department of Engineering Science at the University of Oxford and the North West Control Engineering Group of ICI at Runcorn, which makes it possible to do experimental work on full-scale industrial processes. The chapter describes investigation using a dedicated PDP 11/34 computer interfaced so as to provide feedback control of the pH of effluent at one particular ICI plant. The computer replaces a conventional analogue PID controller in a typical single-input single-output pH-control loop where there appeared to be scope for improving performance.

The computer was installed early in 1978, and investigations since then have been staged as follows:
(a) Implement conventional digital PID control to verify the installation and to demonstrate that the computer can perform at least as well as the analogue controller. This was completed in February 1978 and is not discussed further here.
(b) Imbed the PID control law in a jacketing algorithm which takes

account of what is known about process nonlinearities and of available feedback and feedforward signals. At this stage the jacketing algorithm is based on well-established concepts of linearising by means of an assumed inverse pH-characteristic, of cascade control of reagent flow, and of feedforward control. What is novel is the ease with which the algorithm can be developed and implemented using the online computer; this stage was completed in June 1978.

(c) Retain the jacketing algorithm of stage b but replace the PID control law by recently-developed general purpose 'adaptive' control laws such as self-tuning control [3,4] or optimal k-step-ahead control [5,6].

(d) Use recently-developed concepts [7] of dynamic state, estimation and optimal control to account for all available information, for example using an iterated extended Kalman filter, so as to generate control which may be close to optimal. The nonlinearity of eqn. (8.1) places theoretically optimal pH control beyond the bounds of practibility. Performance at each stage is monitored by the computer and by plant-operating personnel. The computer produces time histories of significant process variables and calculates integrated performance criteria based on both quality and cost of neutralisation. The plant operators have a veto on whether or not the computer may continue to control the process; at any time they can switch back to analogue control.

The project is thus structured to bring together aspects of research on pH control which have hitherto appeared piecemeal. The need to use feedforward signals is recognised in References 8–11, linearising functions have been discussed in References 9, 11, and 12–17, and adaptive schemes tailored to specific pH problems have been described in References 13 and 17–19. Modern control technology has been discussed in the form of time-optimal control, [20] and of the state-space approach to questions of controller design [9,21] and stability [14,21,22] and of online digital computing [16,17]. The novel feature of our collaborative project is that it deploys resources, the online computer interfaced to a full-scale production process and programmed in the light of recent theoretical insights, which have not simultaneously been available to other investigators.

This chapter gives a summary of the project, further details can be found elsewhere [23]. Section 8.2 describes the controlled process, Section 8.3 describes the computer installation, and Section 8.4 reports on investigations in stages b and c. The stage b investigations show that the online computer, implementing PID control within a jacketing algorithm based on conventional concepts performs substantially better than the conventional analogue controller. There is nevertheless scope for further improvement, and stage c investigations into whether this can be achieved by general-purpose 'adaptive' controls are still in progress. Special algorithms to perform estimation and control for stage d investigations have been developed, but little experimental work has yet been done. The conclusion (Section 8.5) of the chapter is that modern control technology can contribute substantially to pH control.

8.2 PROCESS

The process investigated was neutralisation of aqueous acidic plant effluent in a continuous stirred-tank reactor (CSTR.) of volume (V) about 15m^3. Under normal production conditions the acid to be neutralised (HCl) was strong, and buffering effects were not significant. Fig. 8.1 shows the principal features of the process and its control systems. Acid load in the form of raw effluent enters the CSTR. from two sources, A and B. Source A (Fig. 8.1) is local, and for most of the time produces a slowly fluctuating load of acid; source B is about 1km distant from CSTR. and is a batch process producing fairly constant acid loading for periods up to five hours, separated by severe transients. Table 8.1 shows the proportions in which the two flows F_A and F_B contribute to the total load ($\bar{F}_A$ and $\bar{F}_B$ are mean values and $F_1 = F_A + F_B$).

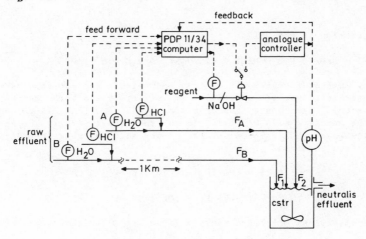

Fig. 8.1 pH-control system

Table 8.1: Components of acid load

	A	B
Mean proportion of total flow	$\bar{F}_A = 0.5F_1$	$\bar{F}_B = 0.5F_1$
Variation in component flow	$0-1.1\bar{F}_A$	$0.85\bar{F}_B-1.15\bar{F}_B$
Variation in acid concentration (kg mol m^{-3})	1-3	0-1

The load is neutralised by a strong aqueous reagent (NaOH) which enters the tank at a rate determined by the position of a valve. The position of this valve is the primary control input affecting neutralisation; it requires continuous adjustment, under feedback control, in order to achieve satisfactory results. The primary feedback signal is pH measured at either of two pH probes in the CSTR and automatically compensated for temperature; it can be in error for various reasons, such as fouling of the probe surfaces, poor mixing in the tank and buffering due to unexpected components of raw effluent, calibration drifts, or other noises.

When the computer was installed, feedforward signals were made available indicating the flows of water and of acid at each source. The acid signals are sometimes in error when purging gases are present at the sources. The feedforward signals from source B refer to loading which has to travel through about 1km of pipework before entering the CSTR. A feedback signal indicating reagent flow was also made available. This was used to give a cascaded flow-control loop at the computer ouput so that the computer could be regarded as adjusting reagent flow, rather than valve position, and effects of valve nonlinearities were made negligible.

The reaction in the CSTR.:

$$HCl + NaOH + H_2O \rightarrow NaCl + Cl^- + Na^+ + H_2O + H^+ + OH^- \qquad (8.2)$$

and the extent to which effluent is neutralised are governed mainly by the concentrations of reagents. If mixing were both perfect and instantaneous the ionic concentrations $[Cl^-]$ and $[NA^+]$ in the outflow from the tank would be related to the total flows F_1, F_2 and concentrations $[HCl]$, $[NaOH]$ of the strong acid and base entering it by conservation equations

$$V\frac{d}{dt}[Cl^-] = [HCl]F_1 - [Cl^-](F_1+F_2) \qquad (8.3a)$$

$$V\frac{d}{dt}[Na^+] = [NaOH]F_2 - [Na^+](F_1 + F_2) \qquad (8.3b)$$

The concentrations must also satisfy an equation of electroneutrality

$$[Na^+] + [H^+] = [Cl^-] + [OH^-] \qquad (8.4)$$

which, together with the dissociation equation for water:

$$[H^+][OH^-] = k_\omega \qquad (8.5)$$

relates them to $[H^+]$ and thus to pH. This relationship is conveniently [16] expressed in terms of ionic concentration difference X defined by

$$X \equiv [OH^-] - [H^+] \qquad (8.6)$$

which combines with eqn. 8.4 to give

$$X = [Na^+] - [Cl^-] \qquad (8.7)$$

and with eqn. 8.5 to give

$$[H^+] = \frac{X}{2}(\sqrt{1 + 4k_\omega/X^2} - 1) \qquad \text{if } X > 0 \qquad (8.8a)$$

$$[H^+] = -\frac{X}{2}(\sqrt{1 + 4k_\omega/X^2} + 1) \qquad \text{if } X < 0 \qquad (8.8b)$$

Fig. 8.2 sketches the relationship of eqns. 8.8a and b, and Fig. 8.3

shows how eqn. 8.1 transforms it into the severely non—linear titration curve relating pH, which can be measured, to ionic concentration difference X, which is what can be controlled by regulating the flow F_2 of reagent so as to change $[Na^+]$.

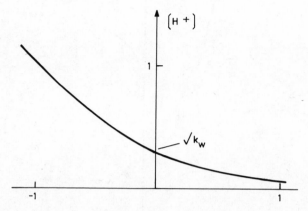

Fig 8.2 Relationship between X and $[H^+]$ (eqn. 8.8)

A simple equation describing the dynamics of the controlled process is obtained by subtracting eqn. 8.3a from eqn. 8.3b and using eqn. 8.7 to give

$$X + \frac{V}{F_1 + F_2} \frac{dX}{dt} = \frac{F_2}{F_1 + F_2} [NaOH] - \frac{F_1}{F_1 + F_2} [HCl] \qquad (8.9)$$

Because the reagent is concentrated its flow F_2 is always much smaller than the total flow F_1 of raw effluent, so there is little error in writing

$$F_1 + F_2 \approx F_1 \qquad (8.10)$$

and eqn. 8.9 becomes

$$X + T \frac{dX}{dt} = KF_2 - [HCl] \qquad (8.11)$$

where

$$T \equiv V/F_1 \quad \text{and} \quad K \equiv [NaOH]/F_1$$

are, respectively, a time constant and a gain, which vary with the flow F_1 of raw effluent. A more accurate description of CSTR dynamics is given by introducing a dead time τ to account for mixing delays. eqn. 8.11 then becomes

$$X(t+\tau) + T \frac{dX(t+\tau)}{dt} = KF_2 - [HCl] \qquad (8.12)$$

The value of $\tau(=0.05T)$ was found experimentally and was smaller than expected. [9]

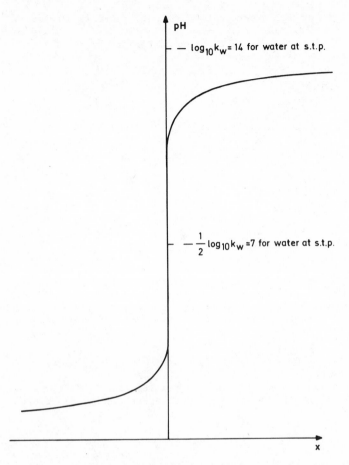

Fig. 8.3 pH characteristic

Fig. 8.4 shows how well the actual response of ionic concentration difference X (computed from measured pH using eqns. 8.1 and 8.8) to a step change in reagent flow F_2 is approximated by the exponential response predicted by eqn. 8.12.

Eqns. 8.1, 8.8 and 8.12 provide a simple mathematical model capturing the main features of the controlled process. They are summarised in the block diagram of Fig. 8.5 which shows linear dynamics affected by the flow F_1 and followed by the severe nonlinearity of pH measurement. The most significant features neglected here are biases and noises on the pH measurement and on the feedforward signals, uncertainty about the timing of feedforward signals due to flow in the pipework from the remote source B, and unexpected additions to plant effluent which would invalidate eqns. 8.2, 8.4, 8.7, and 8.8 by buffering or other reactions.

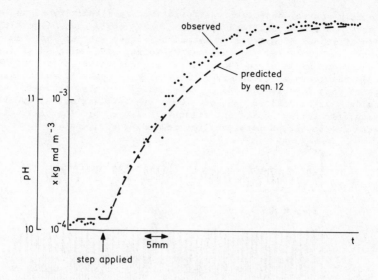

Fig. 8.4 Continuous stirred-tank reactor step response

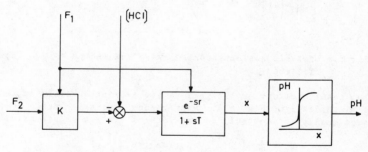

Fig. 8.5 Block diagram of neutralisation process

Other practical features were less significant from the point of view of our investigations. The temperature in the CSTR could vary over the range 0–70°C and cause significant changes in the dissociation constant k_ω of eqn. 8.5. A temperature signal was therefore made available to the computer, which performs more accurate correction of raw pH to s.t.p. than was possible with the original analogue hardware. With this automatic compensation the value of the dissociation constant k_ω in eqn. 8.5 and 8.8 can be assumed [1] to be

$$k_\omega = 10^{-14} \qquad (8.13)$$

The response times of the instruments measuring pH and of the reagent flow-control loop were both of order 10s, which is small compared to T and to τ. A small reagent-flow valve was installed in parallel with the original valve, and the flow-control algorithm included scheduling of the two parallel valves to give improved rangeability.

The main control objective was to achieve an alkalinity of pH = 10.2 or more in the neutralised effluent, which would then merge with other acidic effluents downstream. A second objective was to use as little reagent as possible. The analogue PID controller, with pH as its input and positioning signal to the original reagent valve as its output, was usually unable to maintain stable control below pH = 12 and so used more reagent than necessary. Fig. 8.6 indicates its typical unstable behaviour with desired pH of 10.2; the full extent of excursions, particularly to low pH, is not evident bacause of low-pass filtering effects due to the pH meter and the recording equipment.

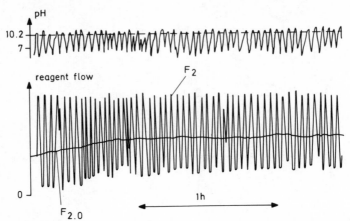

Fig. 8.6 Unstable analogue control with desired pH = 10.2 (low-pass filtered)

The results of our investigations are presented mainly in the form of time histories like Fig. 8.6. Performance criteria were computed to measure quality of control, in the form of average excess acidity

$$I_1 = \frac{1}{T_1} \int_{t_1}^{t_1+T_1} \text{function } [10.2 - \text{pH}] \, dt \qquad (8.14)$$

where function [x] = x if x > 0
              0 if x < 0

and to measure cost of control in the form of average reagent flow

$$I_2 = \frac{1}{T_1} \int_{t_1}^{t_1+T_1} F_2 \, dt \qquad (8.15)$$

The numerical values of these criteria depends strongly on operating conditions which vary from day to day, as well as on what control algorithm is being used. The criteria are therefore less useful as a basis for detailed comparative studies than as a general indication of the effectiveness of control during the period of the investigation.

8.3 COMPUTER INSTALLATION

A PDP 11/34 computer having 32K of core and equipped with a Media [24] interface handling 4-20mA analogue signals was installed in a small room adjacent to the main control room of the plant. Control signals to the two reagent-flow valves were converted to 3-15lb/in^2 pneumatic signals for transmission to the control room where plant operators could choose between computer control and the original analogue control. Software, chosen for purposes other than those of our investigation, was based on ICI's process control package [25] under MTS 2G operating system [26] written in RTL 2 [27] and compiled offline. The computer was controlled from a teletype, and programmes were loaded from fast paper-tape reader. Output from the computer was in the form of analogue signals to a three-channel chart recorder, and in the form of paper tape, from a high-speed punch, for subsequent analysis offline.

Two separate discrete-time sample rates, one for measurement and one for control, could be selected from the teletype. The sample intervals available for measurement were 2^m seconds ($m = 0,1,2,...$) and control action could be implemented every 2^n samples ($n = 0,1,2,...$). The faster measurement rate was used to produce smoothed measurement at the control sample times.

The computer was adequate in both size and speed for all algorithms investigated. In stage b investigations, PID control with a jacketing algorithm was implemented at sample intervals down to 4s. In preliminary stage d investigations, an iterated extended Kalman filter with eight state variables and five iterations was implemented at sample intervals down to 8s. These sample intervals are small compared to the process time constants T and τ.

8.4 INVESTIGATIONS

8.4.1 Introduction

The simple model in Section 8.2 of the controlled process contains information which can be exploited in a control algorithm. If the process actually satisfies eqns. 8.1, 8.8, and 8.12 it would be possible to linearise the pH nonlinearity and to use feedforward to compensate for changes in the acid load to be neutralised. The necessary calculations are easily implemented by an online digital computer and provide a 'jacket' within which feedback can be used, as shown in the block diagram of Fig. 8.7, to compensate for features neglected in the simple model. The dotted lines in the Figure represent uncertain signals affected by neglected biases and noises and timing problems. Jacketed feedback control uses more information about the controlled process than does plain feedback control, such as is implemented by a conventional analogue controller, and so can be expected to give better performance to the extent that the information is correct. The algorithms investigated in stages b and c were jacketed feedback controls. The algorithm of stage d takes account of additional information about features neglected in the simple model, and so, at the cost of greater complexity, may be able to give further improvements in performance.

The linearising calculation for jacketed feedback control combines eqns. 8.1 and 8.8 to give the error:

$$e \equiv X_o = X$$

between ionic concentration difference X_o, corresponding to the desired value pH_o of pH, and X corresponding to measured pH.

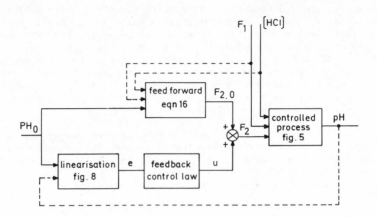

Fig. 8.7 Block diagram of jacketed feedback control

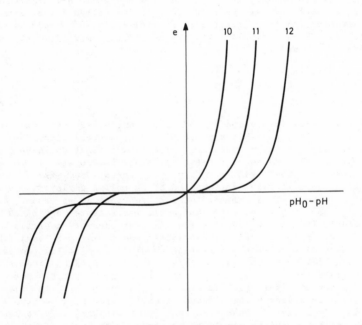

Fig. 8.8 Linearising function

Fig. 8.8 shows the relationship between e and error in pH for several values of pH_0. The feedforward calculation is based on a steady-state ($dX/dt = 0$) version of eqns. 8.11 or 8.12 which gives the steady-state value $F_{2,0}$ of reagent flow to achieve a desired X_0

$$F_{2,0} = ([HCl] + X_0)F_1/[NaOH] \qquad (8.16)$$

The value of F_1 in eqn. 8.16 is the sum of measured flow F_A from the local source and a delayed value of flow F_B from a remote source, the amount of delay being related to the flow F_B. The actual reagent flow in jacketed feedback is then

$$F_2 = F_{2,0} + u \qquad (8.17)$$

where u is generated by the feedback control law as shown in Fig. 8.7.

8.4.2 Jacketed PID control (stage b)

The first jacketed feedback control law to be investigated was conventional PID control with integral desaturation. In the notation of Fig. 8.7,

$$u(i) = K_c(e(i) + K_I Q(i) + K_D(e(i) - e(i-1))/\Delta) \qquad (8.18a)$$

where i is an integer representing discrete-time sample points, K_c, K_I and K_D are parameters of the controller, Δ is the control sample interval and $Q(i)$ is the desaturated

$$Q(i) = Q(i-1) + e(i)\Delta \qquad \text{if } 0 < F_2(i) < F_{2,max} \qquad (8.18b)$$

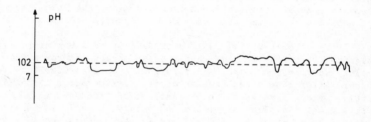

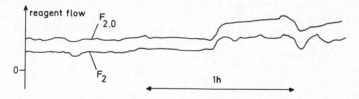

Fig. 8.9 Jacketed PID control with desired pH = 10.2

This algorithm could be tuned to control the process at values of pH down to values as low as the desired value 10.2. Fig. 8.9 is a typical record of the algorithm's performance; it compares favourably with the performance of the analogue controller shown in Fig. 8.6. The feedforward reagent flow $F_{2,0}$ shown in the Figure indicates the loading on the neutralisation process; the conditions shown are severe in both magnitude and rate of load changes.

Jacketed PID control has become the normal algorithm for computer control when other investigations are not being made. Its superiority over conventional analogue control is confirmed by the plant operators, who have allowed it to run unsupervised for extended periods of several days, and by plant management, who are seeking to retain the computer for routine control purposes. The superiority is mainly owing to the computer's ability to exploit information, but its greater accuracy in implementing small values of the control gain K_c is also a factor.

Although superior, the PID algorithm was similar to the analogue controller in that its performance could not be sustained unless the parameters K_c, K_I and K_D were manually retuned from time to time. The need to retune would be indicated by poor performance and is attributed to changes in neglected factors affecting the uncertain signals shown dotted in Fig. 8.7, especially to bias on the remote source B. To meet this type of need automatically is the objective of 'adaptive' control, [28] so it was natural that the next stage of investigation should be to replace the PID control law in Fig. 8.7 by adaptive control laws.

8.4.3 Jacketed adaptive control (stage c)

Two adaptive control algorithms are currently under investigation. Self-tuning control (STC) [4], which is well developed and simple in that it uses a certainty-equivalent [29] control law, is implemented in the online computer and has been investigated on the full-scale process. Optimal-k-step-ahead (OK) control which, although less well developed and the subject of current research, [5,6] is likely to give better performance than STC because it is cautious, [30] has been the subject of a simulation study and will soon be implemented online.

The STC algorithm [4] was provided by P.J.Gawthrop. It uses exponential forgetting to give continuous adaption, incremental ouput to give integral action, and model following to give stable closed-loop behaviour. After some manual tuning of its parameters, this algorithm was able, under favourable conditions, to control down to pH = 10.2 as effectively as the PID algorithm of stage b and to sustain its performance by adapting to slow changes in the process and to artificial steps in pH bias. Rapid load changes or off-scale pH measurements, which are not unusual, were not favourable to STC performance. They could each cause the algorithm to mistake the sign of the controlled process gain and to give unacceptable performance. Incidence of such performance was minimised by imbedding the STC algorithm in a 'corset' which transferred feedback control to a PID algorithm whenever measured pH was outside an allowable range. Fig. 8.10 shows typical behaviour of the corseted STC in the presence of a major load change.

On average, the STC did not appear to be clearly superior to PID control. Under favourable conditions it could sustain performance as good as the best PID without manual retuning, but under other conditions, when the corset was in action, its performance was worse. It has been run for extended periods, but the plant operators are inclined to switch from it back to analogue control when its control action appears to them to be counteractive. The possibility of improving STC performance by further manual tuning of its parameters is to be the subject of a simulation study.

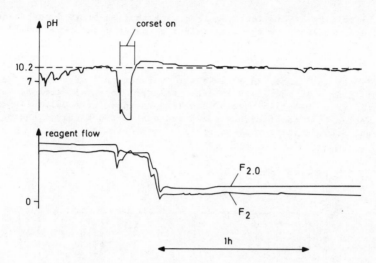

Fig. 8.10 Jacketed corseted STC control with desired pH = 10.2

The OK algorithm investigated here uses a basic version of OK control, known [6] to compare favourably with STC, modified to include integral action and model following. The algorithm assumes that it is controlling a process which satisfies a first-order difference equation

$$e(i + 1) + ae(i) = bu(i) + \xi(i + 1) + c\xi(i) + d \qquad (8.19)$$

in which a,b,c and d are unknown stochastic coefficients, and $\xi(i)$ is an independent white noise of known variance σ. The algorithm is simple because eqn. 8.19 is only of first order and has time delay (k of 'optimal-k-step-ahead') of only one sample interval (k = 1). It is very unlikely that the jacketed pH process would actually satisfy an equation as simple as eqn. 8.19, but the simplifying assumptions here are no more extreme than those implicit in the use of PID control, which could only be optimal for a controlled process of very low order. The coefficients a,b,c and d are assumed to be characterised by further first-order stochastic equations of the form

$$x_j(i + 1) = g_j x_j(i) + \alpha_j + \xi_j(i) \qquad j = 1,2,3,4 \qquad (8.20)$$

where x_j is the stochastic variable and g_j, α_j and the variance σ_j of the white noise ξ_j are known constants. The mean and variance of x_j are related to these constants by

$$\bar{x}_j = \alpha_j/(1 - g_i) \qquad (8.21a)$$

$$\text{var }(x_j) = \sigma_j/(1 - g_j^2) \qquad (8.21b)$$

and so the g_j, α_j and σ_j can be chosen to match what is known about the means and variances and rates of change of the coefficients of eqn. 8.19. The stochastic coefficients are estimated by an extended Kalman filter which has been found here and elsewhere [31,32] to have good convergence properties.

The resulting estimates, together with their estimated covariance, are used by a control law which minimises a k-step-ahead (k=1) cost function

$$J = E[(e(i + 1) - pe(i))^2]$$ (8.22)

Modifications here to the basic version of OK control are the coefficient d in eqn. 8.19, which includes integral action, and the second term pe(i) in eqn. 8.22 which provides for model following. The assumption that coefficients a,b,c and d are stochastic, characterised by eqn. 8.20, ensures that adaptation is continuous.

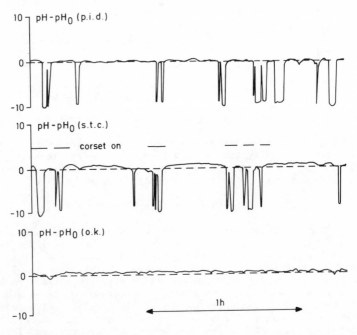

Fig. 8.11 Simulated performance under three jacketed feedback control laws

Because OK control is less well developed than STC a simulation study was undertaken to compare its performance with that of STC and of PID as a jacketed feedback control in the configuration of Fig. 8.7. The study was performed on a digital computer programmed to simulate the controlled process of Section 8.2 including stochastic loading, measurement noises, biases and the dynamics of the upstream pipework between source B and the CSTR. The three control algorithms simulated were identical with those implemented in the online computer. Their parameters were tuned to give good simulated performance rather than choosing values identical with those used online. Differences arose because the simulation was not perfect; it was nevertheless thought to have been sufficiently accurate to give useful guidance on the relative merits of the controllers.

Fig. 8.11 shows typical behaviour of the simulated system with each of the three control laws controlling in the face of the same sequence of random variables ξ. The time histories of simulated error between measured PID and STC are similar, as reported above, and that the performance of OK control is superior to both. The question of whether this superiority can be realised in practice is to be investigated shortly.

8.5 CONCLUSIONS

An online digital computer has been interfaced to control the pH of effluent from a full-scale industrial process. It was programmed to implement jacketed feedback control in which a jacketing algorithm performed linearisation and feedforward calculations, while a feedback control law compensated for neglected features. The resulting performance with either PID or STC feedback control was superior to that of the original analogue PID controller: over five months of continuous operation when all three controls were used extensively the average values of the performance criterion I_1 of eqn. 8.14 achieved by the computer were smaller (better) than the average achieved by the analogue controller in the ratios

$$\frac{I_1 \text{ (jacketed PID)}}{I_1 \text{ (analogue PID)}} = 0.5$$

$$\frac{I_1 \text{ (jacketed STC)}}{I_1 \text{ (analogue PID)}} = 0.49$$

Some of the improvement in performance is due to the reagent flow-control loop which eliminated flow-valve non-linearities under computer control but was not available to the analogue controller. However, the improvement is thought to be mainly because of the control algorithms.

Corresponding ratios of the averages I_2 of eqn. 8.15 show that the improved performance was achieved with a slight reduction in the amounts of reagent used:

$$\frac{I_2 \text{ (jacketed PID)}}{I_2 \text{ (analogue PID)}} = 0.97$$

$$\frac{I_2 \text{ (jacketed STC)}}{I_2 \text{ (analogue PID)}} = 0.83$$

The second of these ratios is low (good), partly because the STC control was used mainly when load conditions were not severe; PID control, both analogue and jacketed, were under all these conditions.

It is concluded that an online digital computer can give substantially improved control of pH because of the ease with which it can process information and exploit it by implementing jacketed feedback control. The jacket must be tailored to the controlled process but the feedback control law can be a general-purpose law which only needs to have its parameters tuned to the process. Tailoring and tuning are facilitated if the computer can be programmed online in a high-level language. A simulation study indicates that OK control is a promising candidate for general-purpose feedback control.

Our collaborative project has thus shown that modern control technology can contribute substantially to pH control and, it is conjectured, to other difficult practical control problems. The full extent of the potential contribution is not yet known, so, as it seems unlikely that optimal control theory which addresses this type of question could ever answer it for the nonlinear stochastic pH problem, the experimental project continues. Further stage c investigations will include a simulation study to find out whether the performance of jacketed STC control can be improved by tuning, and trials of OK control on the plant to find out whether its simulated promise is fulfilled . Stage d investigations will continue with a simulation study to find out whether an iterated extended Kalman filter can estimate state variables and biases which are neglected by the feedforward jacket of stage c, and whther the resulting estimates can be used to give better control.

8.6 ACKNOWLEDGMENTS

We are grateful to many people at ICI who cheerfully collaborated with us, particularly the plant operating personnel, and to P.J. Gawthrop who provided the STC algorithm, advised us on tuning it and wrote some of the simulation routines which were used. The work was supported by an SRC CASE award.

8.7 REFERENCES

1. KNEEN, W.R., ROGERS, M.J.W., and SIMPSON, P.: 'Chemistry facts, patterns and principles', (Addison-Wesley, 1972).
2. HMG: Control of Pollution Act, Sections 31 and 34, 1974.
3. ASTROM, K.J., and WITTENMARK, B.: 'On self-tuning regulators', Automatica, 1973, v.9, pp.185-199.
4. CLARKE, D.W., and GAWTHROP, P.J.: 'Self-tuning control', Proc. IEE, 1979, v.126, (6). pp.633-640.
5. JACOBS, O.L.R., and HUGHES, D.J.: Simultaneous identification and control using a neutral control law'. Presented at the 6th IFAC congress, Boston, Aug. 1975, paper 30.2.
6. JACOBS, O.L.R., and SARATCHANDRAN, P.: 'Comparison of adaptive controllers', Automatica, 1980, v.16, pp.89-97.
7. JACOBS, O.L.R.: 'Intoduction to control theory' (OUP, 1974).
8. WILSON, H.S., and WYLUPEK, W.J.: 'Design of pH control systems', Meas. and Control, 1969, v.2, pp.336-342.
9. ROBERTS, P.D.: 'Nonlinear control of a neutralisation process', ibid., 1971, v.4, pp.151-157.
10. SHINSKEY, F.G.: 'pH and PION control in process and waste streams', (Wiley, 1973).
11. IVERSON, A.A., and OCHIAS, S.: 'Controlling effluent pH', Instrum. Tecnol., 1977, v.24, pp.65-68.
12. WALTER, S., and WILKIE, K.: 'Ein neues Verfahren zur pH-Wert-Regelung', Chem. Ing. Tech., 1973, v.45, pp.1071-2.
13. SHINSKEY, F.G.: 'Adaptive pH controller monitors nonlinear process', Control Eng., 1974, v.21, pp.57-59.
14. RANG. E.R.: 'Application of the Popov criterion to pH-neutralisation control', Adv. Instrum., 1975, v.30, pp.764/1-5.
15. BELLETRUTTI, J.J., CHENG, R.M.H., and SLOAN, D.M.: 'Control of industrial acidic waste system subject to large acid influx', ISA Trans, 1976, v.15. pp.274-281.
16. NEIMI, A., and JUTILA, P.: 'pH control by linear algorithm'. Presented at IFAC conference on digital computer applications to

process control, The Hague, June 1977.
17. WOLFENDER, E., BEYERLE, T., and ALT, M.: 'Kontinuierliche Abwasserneutralisation mittels eines Kleinprozessrechners', Regelungstechnische Praxis, 1978, v.20, pp.116–120 and 151–154.
18. PETERS, R.W.: 'Selbstanpassende pH–Wert–Regelung', Regelungstechnische Praxis, 1970, v.12, pp.10–14.
19. GUPTA, S.R., and COUGHANOWR, D.R.: 'Online gain identification of flow process with application to adaptive pH control', AIChE J., 1978. v.24, pp.654–664.
20. McAVOY, T.J.: 'Time optimal and Ziegler–Nichols control', Ind. Eng. Chem. Process Des. Develop., 1972, v.11, pp.71–78.
21. ORAVA, P.J., and NIEMI, A.J.: 'State model and stability analysis of a pH control process', Int. J. Control, 1974, v.20, pp.557–567.
22. KIRLIN, R.L., and MARSHALL, W.C.: 'The case for state variables in pH controller design', Control Eng., 1975, v.22. pp.47–49.
23. HEWKIN, P.F.: 'The control of pH using modern algorithms and online computers'. D. Phil. thesis, University of Oxford, 1979.
24. BROWN, C.M.: 'An introduction to MEDIA and a guide to its maintenance', ICI NWRCEG Report, May 1976.
25. COCKLE, D., and SCRUTTON, D.W.: 'Plant control package for small systems – functional specification'. SPL International Report, April 1977.
26. ICI: 'An introduction to the MTS operating system running on the PDP 11', RTL/2 reference 91, 1974.
27. ICI: 'Introduction to RTL/2', RTL/2 reference 2, 1974.
28. WITTENMARK, B.: 'Stochastic adaptive control methods: a survey', Int. J. Control, 1975. v.21. pp.705–730.
29. PATCHELL. J.W., and JACOBS, O.L.R.: 'Separability, neutrality aand certainty equivalence', ibid., 1971, v.13, pp.337–342.
30. JACOBS, O.L.R., and PATCHELL, J.W.: 'Caution and probing in stochastic control', ibid., 1972, v.16, pp.189–199.
31. SARATCHANDRAN, P.: 'Simultaneous identification and control of discrete–time single–input single–output systems'. D. Phil. thesis, University of Oxford, 1978.
32. LJUNG, L.: 'Asymptotic behaviour of the extended Kalman filter as a parameter estimator for linear systems', IEEE Trans., 1979, AC–24, pp.36–50.

APPENDIX 8.1   LIST OF SYMBOLS

$[\cdot]$	=	molecular or ionic concentration of component
A	=	local source of effluent
a	=	stochastic coefficient (eqn. 19)
B	=	remote source of effluent
b	=	stochastic coefficient (eqn. 19)
CSTR	=	continuous stirred tank reactor
c	=	stochastic coefficient (eqn. 19)
d	=	stochastic coefficient (eqn. 19)
E	=	expectation operator
e	=	error in ionic concentration difference
F_1	=	total flow of effluent to be neutralised
F_2	=	flow of neutralising agent
$F_{2,0}$	=	steady-state value of F_2 to achieve desired pH
$g_j(j=1,2,3,4)$	=	constants specifying rates of change of a,b,c,d(eqn. 20)
I_1	=	average quality of control (eqn. 14)
I_2	=	average cost of control (eqn. 15)
i	=	discrete time (integer)
J	=	cost function for OK control (eqn. 22)
K	=	gain of CSTR($K \equiv [NaOH]/F_1$)
K_C	=	gain of PID controller
K_D	=	derivative time constant of PID controller
K_I	=	integral time constant of PID controller
k	=	time delay assumed for OK control
k_ω	=	dissociation constant of water
OK	=	optimal-k-step-ahead control
PID	=	proportional plus integral plus derivative, three-term control
p	=	constant specifying model to be followed by OK control (eqn. 22)
Q	=	integral term of PID controller (eqn. 18)
STC	=	self-tuning controller
t	=	continuous time
t_1	=	starting time for I_1 and I_2 (eqns. 14 and 15)
T	=	time constant of CSTR ($T \equiv V/F_1$)
T_1	=	averaging time for I_1 and I_2 (eqns. 14 and 15)
u	=	feedback control action
V	=	volume of CSTR
X	=	ionic concentration difference (eqn. 6)
X_0	=	value of X corresponding to desired pH
$a_j(j=1,2,3,4)$	=	constants specifying mean values of a,b,c,d (eqn. 20)
Δ	=	control sample interval
σ	=	variance of ξ
$\sigma_j(j=1,2,3,4)$	=	variance of ξ_j
τ	=	dead time of CSTR
ξ	=	independent random variable (eqn. 19)
$\xi_j(j=1,2,3,4)$	=	independent random variables (eqn. 20)

Chapter 9

Direct digital control in CEGB power stations

9.1 INTRODUCTION

This chapter surveys the application of direct digital control within CEGB fossil-fired power stations. At the present time DDC is only widely applied on boiler controls. Four boilers (at Skelton Grange, Thorpe Marsh, Pembroke and Didcot Power Stations) are under complete digital control, several more have particular plant areas under computer control and many more DDC systems are in the pipeline. This reflects the accelerating trend away from analogue and towards digital control, and significant resources within the CEGB are being allocated to the research into and production of computer control hardware and software.

In this quickly changing environment there are many areas of interest and in this chapter the following topics have been selected for discussion.

The Cost/Benefits of DDC

A commercial organisation such as the CEGB is only prepared to adopt DDC if it is cost effective compared with the alternatives. An attempt is made to specify in what circumstances this is the case.

Scope of the Boiler Controls

A brief description of the main boiler loops under automatic control will be given, and in order to illustrate the complexity of these loops, load and pressure control will be described in more depth.

DDC Hardware

The hardware used in the digital control installation at Thorpe Marsh Power Station will be described to give an indication of modern power station practice.

The CUTLASS Software

The CEGB has developed a software system specifically for on-line computer applications. Part of this software consists of a very high level, block structured language for writing DDC programmes, and this language is specifically designed to be simple enough to be safely and easily used by power station engineers unversed in computer science. The main features of the language will be described.

9.2 THE COST/BENEFITS OF DDC

There are a variety of situations in which the installation of a DDC system might be considered, e.g. for a new power station, for a power station with obsolescent automatic controls, or for a power station with modern analogue electronic controls. In addition the cost/benefit criteria that must be satisfied will vary depending on the planned plant useage and on the economic climate. For example, during the current recession there is a low demand for electricity which has resulted in transient spare generating capacity and tight cash constraints. The planned closure dates for old, small stations have been brought forward, and the newer, large units are being required to operate more flexibly. Thus the benefits of control refurbishment projects have tended to decrease in these small stations and increase in the large ones. However the possible benefits of DDC are large since a modern power station spends enormous sums of money on fuel and hence significant savings can be made from small percentage improvements in the efficiency and availability of the plant. For example, for Drax, which currently has 3 660MW boiler-turbine units, the coal bill in 1979 was around 150 million pounds and this cost is directly related to conversion efficiency. In addition, if a Drax unit is unavailable the power must be generated by a less efficient unit at an additional cost of the order of 60,000 pounds per day.

Computer-based control and instrumentation systems contribute to increased unit efficiency and availability in the following ways:-

9.2.1 Reliable Information Displays

Unit operators are required to diagnose fault conditions and take corrective action quickly so that the possibility of plant damage and loss of availability is minimised. To do this effectively requires reliable display of the condition of the plantN and in any major control refurbishment, whether by analogue or digital equipment, the increased reliability of instrumentation is often the greatest benefit. Computer-based systems, however, allow the cross checking of process measurements and the sophisticated manipulation of information to generate highly meaningful displays. Unfortunately, the benefits of such displays are difficult to measure since incidents in which there is major plant damage are rare and it is often difficult to ascertain the cause of the fault. However, the cost of such incidents can run into millions of pounds.

9.2.2 Less Maintenance

Computer-based control systems tend to be more reliable and easier to maintain than modern electronic analogue equipment - which is itself better than the old pneumatic controllers installed on power stations in the early sixties. Also computer systems have the advantage that they can be used to identify certain plant faults, for example excessive backlash on a valve, and hence lead to better direction of maintenance effort. In addition several changes in control strategy are likely over the life of a power station as understanding of its performance develops or as new instrumentation becomes available. Such 'maintenance' of the

control structures is much easier with a software programmed DDC system
compared with a hardwired analogue one.

Some of these benefits may be demonstrated by results obtained from
Thorpe Marsh where a distributed DDC scheme replaced the obsolescent
pneumatic control system. Prior to the refurbishment 55 emergency
maintenance cards were issued per month on the milling plant, whereas
after the refurbishment the rate fell to an average of 12 per month
(with a proportionate saving in manpower costs). Since these cards are
generally only issued when a control system fault is causing loss of
generation, one would expect improvements in availability. Indeed, an
analysis of the causes of loss of generation showed that, prior to the
refurbishment, the loss of generation due to mill unavailability had a
replacement cost of 24,000 pounds per month on average, whereas
afterwards the cost was 4,000 pounds. While some of this saving may be
due to other causes, it has been observed that the improved
instrumentation has led to earlier and more reliable diagnosis of faults
in the mechanical and electrical components of the milling plant.

9.2.3 Improved Control

Boilers are highly interactive systems, and transients in one input
tend to affect many plant variables. Good automatic controls minimise
these interactions and increase the likelihood of the unit continuing in
operation following a fault condition. For example, tight drum level
control is required to keep the unit operational if, say as a result of
severe load rejection, the superheater safety valves blow. In addition,
good automatic controls keep units operating close to optimum efficiency
despite the continuously fluctuating consumer demands placed on the
electricity supply system.

The savings due to efficiency can be demonstrated from observation of
the performance of the Thorpe Marsh superheater temperature controls.
Here it is important to avoid overspraying, otherwise very costly damage
to the turbine can occur. When no automatic controls were available on
superheater outlet steam temperature there were large fluctuations in
temperature, which required the temperature to be kept below its design
level by, on average, about 4^oC. When an automatic control scheme was
installed, which included overspray protection by computation of the
saturation line of the steam, control was better allowing the controller
desired value to be set at the designed working temperature. The saving
in efficiency due to this is of the order of 25,000 pounds per year and
the computer, plant hardware and manpower cost for this part of the
system was of the order of 75,000 pounds. Further benefits accrue from
reduced maintenance and greater reliability and security.

The examples illustrating the above three benefits make the case for
automatic control, but not necessarily for digital rather than analogue
control. However, if the hardware cost of implementing a number of
similar simple PID control loops is compared for digital and analogue
equipment, it is seen that, as shown in Fig. 9.1, the digital approach
is cheaper for large systems – even if the controls are distributed
within several computers. The price advantage for digital control
becomes greater as the control loops become more complex and involve
adaptive, multivariable or non-linear elements.

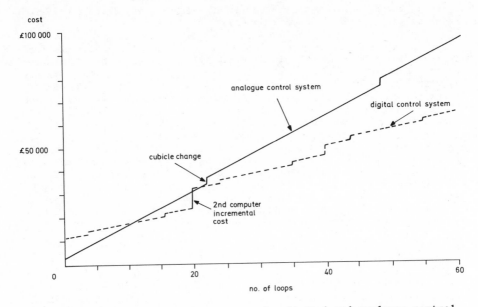

Fig. 9.1 Comparison of hardware costs for digital and analogue control
systems

From this discussion of the benefits of DDC it is clear that an
accurate costing of their value is difficult and extremely time
consuming. However, the benefits are such that DDC is currently being
specified, and is likely to continue to be specified into the
foreseeable future, for new power stations. It is also likely to be used
for major refurbishments of the controls on existing stations if these
become obsolete or too unreliable. However the benefits are not yet
generally considered to be such that, on a wide scale, existing
electronic analogue control systems that function as originally designed
are being taken out and replaced by more sophisticated or more reliable
computer based systems. As yet the ability of the computer to diagnose
faults and the advantages of modern algorithms over simple PID control
are only being realised on pilot projects for small areas of plant.

9.3 SCOPE OF THE BOILER CONTROLS

On a coal-fired drum boiler in a power station there are four main
control areas each of which involve a number of control loops. These
areas are:-

9.3.1 Load and Pressure Control

The primary purpose of a power station is to generate electrical power
or 'load'. In a coal-fired boiler, coal is burnt to generate heat, and
this heat is used to boil water. Increasing the heat input increases the
boiler steam pressure, and this in turn leads to increased steam flow
from the boiler and through the turbine, and hence to increased load. At
the same time steam flow may be regulated by modulation of the turbine

governor valves. However, closure of the valves reduces steam flow but causes the boiler pressure to build up. Thus there is an intimate relationship between load and pressure and hence between their controls.

9.3.2 Feedwater Control

In a boiler, water is fed in, heated up until it evaporates, and then this steam is superheated. The boiling water/steam boundary occurs in the drum, and the position of this boundary must be held in the centre of the drum to avoid damage to boiler tubes. To do this the feed flow must be varied (normally by opening or closing feed valves) so that it matches the varying steam flow requirements. At the same time cavitation in the feed pumps must be avoided.

9.3.3 Steam Temperature Control

As described in 9.3.1, the steam pressure depends on the total energy input and ouput flows from the boiler. However, the steam temperature depends on the distribution of the input heat flux over the boiler tubes. It is desirable to keep steam temperature under tight control to optimise the compromise between efficiency and tube damage, hence variations in heat flux distribution are compensated for either by spraying water into the superheater tubes, or by adjusting dampers which change the distribution of the gas flow.

9.3.4 Combustion Air Controls

To vary the heat input to the boiler and hence the power generated, the fuel supply to the boiler must be varied. In coal-fired boilers, coal is ground to a fine dust and pneumatically conveyed in a rich (and hence non-combustible) mixture to the burners. Here extra air is added to form a highly combustible mixture which is blown into the furnace. To optimise the efficiency of the combustion process, continuous adjustments must be made in the air supply to the boiler to follow the variations in fuel supply and maintain a stochiometric air/fuel ratio in the furnace. In addition, the pressure in the furnace must be maintained just below atmospheric so that the noxious exhaust gases are sent through gas cleaners and up the stack and do not leak out of holes in the furnace walls.

From these descriptions it should be clear that there are benefits in achieving good dynamic response from the boiler controls so that they may react quickly to the varying demands placed on the plant. Since the plant dynamics are also highly interactive, the control structures tend to be complex. This contrasts with the status of control in petro-chemical plants where the benefits come largely from optimisation of steady state performance. To illustrate further the complexity of the controls, a detailed description of possible load and pressure control structures will now be given.

9.4 STRUCTURES FOR LOAD AND PRESSURE CONTROL

The CEGB has a statutory requirement to maintain the grid frequency within tight limits and, since frequency depends directly on the kinetic energy stored in all the rotating machinary on the system, to do this the total generated power must be kept in balance with the total consumed power. However, large, fast and unpredictable changes can occur in both generation and consumption and hence, to maintain this balance, every boiler-turbine unit is fitted with a high gain proportional control loop from frequency error to turbine governor valve position and thus to generated load. This loop causes errors between generation and consumption to be corrected within seconds, but this is at the expense of the steady operation of the boiler plant.

Some of the fluctuations in power demand, such as those occurring at the end of a popular TV programme, can be predicted and the most economic strategy of generation can be planned in advance. To allow for this it is desirable that every boiler-turbine unit should be able to change its generated load at the request of Grid Control at a rate of up to 5% of full load per minute. Errors in meeting such requests contribute to the unpredictable component of generation and are ultimately corrected by the tight frequency control loop. Therefore, in responding to these requests, disturbances of the plant operation that incur costs in the form of reduced efficiency or plant life are to be avoided.

9.4.1 Plant Interactions

The load generated by a turbogenerator is an approximately linear function of its inlet steam pressure. For a drum boiler, the turbine inlet pressure depends on the steam pressure in the drum, the pressure drop from drum to superheater and the pressure drop across the turbine governor valve (TGV). The drum pressure, which for small changes, is linearly related to the stored energy of the boiler, responds to the imbalance of energy flows into and out of the boiler. The input energy flow can be manipulated by the demand sent to the firing system, and the output energy flow by the TGV position demand. The strong interaction between load and boiler pressure is clear in the linearised approximate transfer function matrix of the system which is:-

$$
\begin{bmatrix} P_{DR} \\ P_S \\ L \end{bmatrix} = \frac{1}{(1+Ts)} \begin{bmatrix} MQ(s) & M \\ (A-B)Q(s) & -(A+BTs) \\ G(s)\,Q(s) & G(s)Ts \end{bmatrix} \begin{bmatrix} Q_D \\ V_D \end{bmatrix} \tag{9.1}
$$

where:-

P_{DR} = drum pressure

P_S = superheater outlet pressure

L = generated load

Q_D = firing demand

V_D = TVG position demand

$Q(s)$ = firing system transfer function

$G(s)$ = turbogenerator transfer function

T = boiler time constant

A,B,M = boiler parameters

Because of this interaction, integrated design of the load and boiler pressure controls is required. Generally, but not exclusively, pressure control is based on superheater outlet pressure and the performance criteria that should be satisfied by the combined controls, are listed below in order of importance.

(i) Boiler pressure transients must be kept within such limits that the unit remains operational even during severe frequency disturbances. Only in most exceptional circumstances should disturbance cause a unit trip.

(ii) A fast and well defined load response must be made to frequency transients.

(iii) Boiler pressure should be maintained close to its desired value as this eases boiler operation.

(iv) The unit should be able to run at steady load, and change load at a specified average rate at the request of Grid Control.

9.4.2 Boiler – Follows – Turbine Control

The control systems can be structured in a number of ways to achieve these criteria. The conventional approach, termed Boiler-Follows-Turbine, is to implement load and frequency control via the TGV and pressure control via the fuel. To avoid conflict between the load and frequency controls, the load change required by frequency may be applied as an offset on the load desired value. The transfer function matrix of the controller is:–

$$\begin{bmatrix} Q_D \\ V_D \end{bmatrix} = \begin{bmatrix} K_{11}(s) & 0 & 0 \\ 0 & K_{22}(s) & R_V(1 + K_{22}(s)) \end{bmatrix} \begin{bmatrix} E_p \\ E_L \\ E_f \end{bmatrix} \quad (9.2)$$

where $K_{ij}(s)$ represents a proportional–plus–integral controller transfer function, E_p, E_L, E_f represent the errors between desired and measured values of pressure P_S, load L, and frequency f, and R_V represents the gain from frequency to the TGV. The advantages of this structure are, firstly, that it is simple and gives a predictable response to frequency, and secondly, the pressure is only disturbed by the necessary amount during frequency transients. The disadvantage is that pressure is unduly sensitive to firing disturbances. This is particularly relevant if firing becomes limited during rapid changes in desired load, in which case severe fluctuations in pressure will occur.

9.4.3 Turbine – Follows – Boiler Control

An alternative structure, termed Turbine-Follows-Boiler (Waddington, 1979) [1], uses the TGV to control pressure and the fuel to control load. A fast control response to frequency must be maintained and hence must be implemented via the TGV. This is done by transiently altering the pressure set point with a high pass filter, $R_p(s)$, and at the same time changing the desired value of load. The control system has a transfer function matrix:–

$$\begin{bmatrix} Q_D \\ V_D \end{bmatrix} = \begin{bmatrix} 0 & K_{12}(s) & R_V K_{12}(s) \\ K_{22}(s) & 0 & R_V - R_p(s)K_{21}(s) \end{bmatrix} \begin{bmatrix} E_p \\ E_L \\ E_f \end{bmatrix} \quad (9.3)$$

The main advantage of this structure is that the pressure control
remains good despite mill limitations, however, there are two
disadvantages. Firstly, control of pressure is lost when the governor
valve becomes fully open (at full load). Secondly, it is difficult to
determine the superheater pressure change, and hence the form of $R_P(s)$,
required to give the correct load response, as this depends on the
boiler parameters A and B and the transfer function Q(s) of the mills,
all of which are poorly defined.

To ease design of the pressure offset filter, $R_P(s)$, the pressure
control can be based on a combination of drum and superheater pressure,
rather than superheater pressure alone as is normal. A signal of the
form:−

$$P_C = \frac{1}{1+sT_1}\ P_S\ +\ \frac{sT_1}{1+sT_1}\ P_{DR}\qquad(9.4)$$

is generated which has little high frequency response to TGV movements,
and this is used in the pressure control. A rapid movement of the TGV
due to a grid frequency change is not immediately counteracted by the
pressure control, and transient offset of the pressure set point is only
required to shape the mid frequency response to frequency disturbances.

9.4.4 High Security Control

A third type of system, termed High Security (Bransby et al., 1979)
[2], has been developed using modern multivariable analysis techniques.
Pressure control is via both the TGV and the fuel, and load control is
via the fuel alone. This means that pressure remains controlled if
either the TGV or firing system limits, and in normal operation load
control is good. Grid frequency control is implemented using a novel
adaptive structure illustrated in Fig. 9.2.

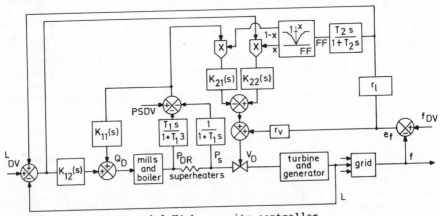

Fig. 9.2 High security controller

The pressure control is based on the combined signal P_C, defined in
eqn. 9.4 and no offset to the pressure set point is made. This gives a
reasonable response for small frequency disturbances. For large
disturbances there is progressive transient switching to load control
via the TGV, so that a predictable load response is obtained despite

limitations in the firing system. It overcomes the main disadvantages of the Boiler-Follows-Turbine and Turbine-Follows-Boiler structures, though it is rather more complex. This is not a limitation when computer control is used.

9.5 ENGINEERING A DDC SYSTEM

At present with rapid changes occurring in computer technology, both the hardware and software used in CEGB digital control projects is changing. However, a brief description of the digital boiler control system commissioned in 1979 at Thorpe Marsh Power Station will give an indication of current practice. This control system was installed on Unit 2 which is rated at 550MW.

The total hardware cost of this project, which controlled 57 loops was in the region of 250,000 pounds which was split as follows:-

7% computers
43% computer interface equipment
13% control desk modifications
19% transducer modifications
18% actuator modifications

The system was totally designed and installed by CEGB staff. Of this manpower effort it is estimated that, like expenditure on hardware, 50% went on development and testing of the computer hardware and software, and 50% on plant modifications.

9.5.1 Distribution of the Control Functions

There are two popular approaches for providing secure digital control systems - namely standby control and distributed control. In standby configurations, large areas of control are implemented in one computer, with a back-up computer to take over in the event of a computer failure. The change-over arrangements tend to be complex and may themselves introduce unreliability. In distributed configurations, the total controls are divided into a number of small areas, each of which can be safely controlled manually by the plant operator. The controls for each of these small areas are implemented in separate computers so that the consequences of a single computer failure are acceptable. More computers are used in distributed systems compared with standby systems, however falling processor costs are tending to make the total cost difference between the approaches small. At Thorpe Marsh, the boiler control functions are distributed over 5 PDP 11/03 computers. Each computer is mounted in its own rack of Media plant interface equipment. The distribution of control function is as follows:-

4 Mill Computer   (13 PID loops)

        4 coal mills
        boiler steam pressure control

5 Mill Computer   (15 PID loops)

        5 coal mills

Combustion Air Computer   (9 PID loops)

>   total air control
>   3 forced draught fans
>   furnace pressure control
>   4 induced draught fans

Steam Temperature Computer   (16 PID loops)

>   4 superheater spray controls
>   2 reheater spray controls
>   4 backpass damper controls
>   4 economiser damper controls

Feedwater Computer   (4 PID loops)

>   4 feed valve controls
>   3 feed pump controls

On the control systems currently being designed, more computers are being used to give wider distribution, and in some cases standby controls are provided on the more important controls such as feed water.

9.5.2 Computers

The computers have between 16 and 24K words of CMOS random access memory each. However, this memory is volatile and its contents will be destroyed if there is interruption to the power supply. Since this would mean the loss of all the control software on the machine, battery back-up of the power supplies to the memory is provided. The processor itself is not backed up and thus loss of automatic control will occur during power supply interruption. The situation is similar with electronic analogue control and hence on all power stations the control power supplies are designed to have a high security. At Skelton Grange Power Station there is a further line of back-up in the form of a 'host' computer which is able to automatically reload the control software from disc, into each of the control computers, if the battery back-up fails to protect the memory. A number of other alternative approaches to obtaining security to power supply transients are possible and these are actively being pursued because such transients are the most common cause of failure of DDC systems.

9.5.3 Interface Equipment

The largest item of expenditure at Thorpe Marsh was on the interface equipment which converted plant signals into numbers accessible to the computer and vice versa. The breakdown of the 536 plant signals used was as follows:-

184 Analogue inputs of which there were
>   47 direct analogue inputs
>   137 relay multiplexed analogue inputs
  5 Analogue outputs
 64 Direct digital inputs
283 Relay isolated digital outputs

The Media interface equipment was designed by ICI and is manufactured by Fisher Controls Ltd., under licence. There are five cubicles into which the Media cards and the computers are mounted, and each cubicle has its own power supplies and is individually wired to the required configuration. The engineering and documentation of this contributes significantly to the cost.

9.5.4 Control Desk

The control desk was originally equipped with many pneumatic gauges, the pressure signals for these being piped from controllers out on plant several hundreds of metres away. These were unreliable and were replaced by modern electrical illuminated indicators. The auto-manual stations were unreliable and unsuitable for connection to the computer. These were replaced by a standardised, CEGB designed unit. As is normal power station practice, the unit generated incremental raise/lower outputs so that it had a 'fail-freeze' facility. The module included push buttons for operator mode selection and colour coded illuminated indication. The new desk layout enabled the plant status to be interpreted at a glance with unusual situations indicated by flashing lights.

9.5.5 Plant Modifications

During the refurbishment many of the obsolete pneumatic transducers on plant were replaced. In addition many of the actuators had to be modified to accept the new computer output drives. In some cases the existing pneumatic positioners were retained, but the split-phase motor which moved the set point arm on the positioner was replaced by a more precise stepper motor. The hydraulic positioners, however, were replaced by a completely new, bi-directional, solenoid valve arrangement [3] (Marsland et al., 1978) in which the positioning was done by the control computer. This concept is being extended further and a device is being developed in which a micro-computer is incorporated in the position measurement transducer and used to generate the drive pulses to position it. Status messages can be then sent back to the control computer over serial line links [4] (Barker et al, 1980). Thus the micro-computer will be able to diagnose faults, such as excessive play in the actuator linkages, and carry out routine maintenance, such as recalibration or resetting of endstops, and inform the operations or maintenance departments as necessary.

9.6 THE CUTLASS SOFTWARE

There is an increasing use of computers within CEGB power stations for all aspects of plant monitoring and control. This is a product of technological change, higher demanded standards, and the requirement for greater efficiency and flexibility. We are at a stage where long term effective use of technological advances is dependent on the ability of engineers to generate, modify, and maintain applications programmes independent of professional software support. In order to achieve this, the CEGB has decided to use standard software for real-time, on-line, power station applications. The standard software adopted is a high-level engineer-oriented system called CUTLASS.

9.6.1 <u>CUTLASS Requirements</u>

Three requirements of the CUTLASS system are especially important:-

(1) The software should be suitable for a wide range of applications within the power station, since the development effort is then spread over a large number of projects. In addition the different applications packages should operate within the same general framework so that future developments can be fitted in without disrupting that which has already been implemented. To achieve this, application language subsets are being developed to handle the computing requirements of the following functions:-

Modulating control
Data logging
Data analysis
Sequence control
Alarm handling
Visual display
History recording

and the CUTLASS framework comprises the main components:-

The TOPSY Real time executive
Communications network management
Support facilities
Integrated I/O and file handling

(2) The software should have a high degree of independence from particular computer types. This is of benefit at the time of purchase as it allows economic advantage to be taken of the choice of available products. Similarly, when it becomes necessary, due to age or obsolescence, to replace the control computers, advantage can be taken of technological developments without incurring large programming costs. This has been achieved by writing CUTLASS in the standard programming language CORAL.

(3) The software should be simple and safe to use so that power station engineers are able to produce and modify programmes with minimal assistance from computer or control specialists. This requirement has been achieved by providing a variety of powerful subroutines within the application languages and by making many organisational and security features transparent to the user.

9.6.2 <u>General Features of CUTLASS</u>

The basic CUTLASS framework supports and is an integral part of the applications packages. Fundamental to this framework is the organisation of programmes and data.

A general requirement in on-line applications programmes is to be able to perform operations, such as take measurements from plant, at regular intervals. In CUTLASS, programme statements are collected together into TASKS each of which is specified to run at a particular repetition rate. The organisation required for this is done by the TOPSY executive and need not concern the user-engineer. Transparent protective features are included to prevent problems from processor overload. TASKS may be collected together into SCHEMES which may be individually put into operation by simple high-level commands. Thus a temperature cascade control scheme may comprise several tasks, such as a valve position loop, a spray control loop and an outlet temperature control loop, each of which runs at a different rate, and all the scheme elements can be ENABLED or DISABLED at the same time. Facilities are also included for running TASKS from interrupts rather than regularly.

CUTLASS data variables and arrays are divided into the 'classes' logical, integer, real, and string and also into the 'types' GLOBAL, COMMON and LOCAL. Local data may only be used in one task, common data may be transferred between schemes, possibly in different computers. Transparent protective rules are built in so that these data transfers are secure and consistent. For example, several tasks should not, in general be allowed to set the value of a particular named variable, as another task using the variable would not 'know' where the value derived from.

A further protective feature of CUTLASS is the concept of 'bad' data. This is generated if, for example, measurements are outside acceptable limits. Bad data propogates through the arithmetic of CUTLASS according to logical rules. Thus, if A and B are real variables with values 3.0 and bad respectively and C is defined as

$$C = A \times B$$

then C will have value bad. Similarly, if A and B are logical variables with values true and bad respectively and

$$C = A \ OR \ B$$

then C will have value true. This distinction between good and bad data greatly increases software security, and, for example, programme execution errors arising from dividing by zero are avoided.

In addition to the programme and data organisation, other general purpose facilities are built into CUTLASS. The communications network management software organises the communication of data between arrays of computers so that the computing functions on plant may be widely distributed. Database management software organises the collection of data from plant at appropriate rates and makes it available to applications programmes as necessary. The DEBUG-MONITOR allows running programmes to be interrogated and their performance monitored during commissioning. Thus, even without the applications language subsets, CUTLASS is a powerful general-purpose on-line computer system.

9.6.3 Features of the Modulating Control Language Subset

In the modulating control subset several of the protective features of CUTLASS are tightened up further to increase security and in fact execution errors are totally avoided. In addition a number of powerful control subroutines, or 'BLOCKS' such as integrators, PID controllers and Rate limits are provided. The blocks are designed to provide the correct function for both the manual and automatic control modes and for the situation in which inputs are bad. The operational mode of all blocks within a task can be controlled by a single statement and a simple structure is provided so that correct initialisation and propogation of modes occurs in complex schemes. Thus the user need only write simple LINK statements to obtain correct function from, for example, a cascade controller with a number of actuators and a variety of switchable control options. In addition only simple procedures need be followed to achieve bumpless manual-auto transfers.

A problem which frequently occurs in control is that of integral (and sometimes proportional and derivative) wind-up of controllers when operational limits are reached. Designing algorithms that respond correctly in such circumstances is one of the most time consuming and error prone aspects of DDC. In CUTLASS however, an automatic, user-transparent method, which makes use of the AUTO-CONSTRAINT operating mode, has been developed. This is extremely easy to use in many situations and is generally safe. However, in particular circumstances its performance can be improved upon and therefore further sophisticated control functions are provided for the expert user such as the INHIBIT BLOCK and the CANNOT function.

Different optimised functions and levels of protection are provided in the other language subsets, and tasks written in different subsets may be mixed in the same scheme. This allows efficient and secure integration of, say, sequence, modulating and alarm software. Thus a particular application might require a programme with 5 or 10 times as many statements and much more programming effort if written in FORTRAN rather than CUTLASS. The protective features are also very important, as the detailed expert analysis that goes into their development need not concern and need not be duplicated by the normal user.

9.7 CONCLUSIONS

This chapter has spanned a range of topics in DDC. In each case of control strategy, hardware and software there is active research and development within the CEGB as advantage is taken of modern theory and technology. One of the primary aims in each of these fields is to produce standard products which can be widely applied so that maximal return can be obtained from limited manpower resources.

9.8 ACKNOWLEDGEMENTS

The work described in this chapter is only in a small part that of the author, and many workers in the Control and Instrumentation Division of the North Eastern Region, Scientific Services Department contributed to it. Particular mention should be made of the CUTLASS development team and of D. Collier and P. Sutherland who were responsible for the Thorpe Marsh refurbishment and the load control work respectively.

This chapter is published with the permission of the Director General of the North Eastern Region of the CEGB.

9.9 REFERENCES

1. WADDINGTON, J.: 'Load Controller Design for a Regulating Coal-Fired Unit', *Proc. IEE*, 1979, v.126, pp.327-332.
2. BRANSBY, M.L., SUTHERLAND, P.: 'Assessment of Load Control Structures

at Drax Power Station', <u>IEE Digest No.1979/21</u>, 1979.
3. MARSLAND, C.R., SLOCOMBE, M.D.: 'Microprocessor Incremental Actuator with Inclination Feedback System', <u>CEGB Technical Disclosure Bulletin No.310</u>, 1978.
4. BARKER, N.J., PRINGLE, S.T.: 'Intelligent Inclinometer', <u>CEGB Technical Disclosure Bulletin No.344</u>, 1980.

Chapter 10

Structured analysis of manufacturing systems

10.1 INTRODUCTION

This chapter comprises two principal parts: firstly, section 10.2 contains a general introduction to the organisation and problems encountered in manufacturing industry, and secondly, section 10.3 contains part of a structured analysis model of a company engaged in batch maufacturing.

Thus, in section 10.2, the concept of a manufacturing system is described in some detail and discussed in terms of the production efficiency and flexibility of such systems. The activities and communications generally found in an engineering manufacturing system are described, as are the principal types of production: these include the methods of job production, batch production, and flow-line production. A brief review of mechanisation and automation is included, and the introduction of numerically-controlled machines is described.

Finally, in section 10.2, reference is made to the <u>off-line processing</u> of data and the <u>on-line control</u> of production processes and information in Computer-Aided Manufacturing Systems, and integrated manufacturing systems are introduced.

The concept of functional modelling of systems through the use of a structured decomposition technique is described briefly in section 10.3, and two types of such model are introduced: firstly, a Functional Model which is concerned principally with <u>activities,</u> and secondly, an Information Model concerned principally with <u>data</u>. This technique, known as Structured Analysis and Design Technique, SADT^{TM}, (SADT^{TM} is a Registered Trade Mark of SofTech Inc.), has been developed by a commercial company, SofTech Inc., Waltham, Massachusetts, MA 02514, USA. The models presented in this chapter are based on this technique but are not intended to be rigorous, being based only on available published literature [1,2,3,4,5]. However they do highlight the principal features of this system which is being used by a number of major organisations engaged in manufacturing.

10.2 MANUFACTURING INDUSTRY

10.2.1 General

This section is intended to be of interest to those involved in the manufacturing industries, that is those concerned with the large-scale, organised production of goods. These industries may be divided into two broad groups: firstly, those involved in the production of materials and commodities, and secondly, those involved in the production of discrete items. By definition, manufacturing industry involves organised activities. The organisation extends to all phases of the activity from the identification of a saleable product and a suitable market, through the actual manufacturing processes, to the sale of the product and the provision of after-sales services, and it involves many extremely complex relationships.

The tasks of creating and maintaining the organisation and of coordinating all phases of the manufacturing activity are those of management. The method, structure and goals of management vary from one concern to another and vary due to changes in economic, social and political climates: the aim may be to maximise profits and ensure high returns on invested capital, to maximise output at any cost or to maintain a stable work load and work force in time of economic uncertainty. In general, however, it may be said that the aim of management in any manufacturing organisation is to deliver goods of a proper quality on time and at an acceptable cost. In order to achieve this it must be ensured that personnel with the appropriate skills are available, along with the necessary tools and materials, in the right place at the right time and all activities must be planned and scheduled, then carefully monitored and controlled to ensure that specific deadlines are met.

10.2.2 Manufacturing Systems

The management of an engineering concern strives to control the interplay between personnel, materials and machines with a view to optimizing the activity as a whole. Traditionally, this control has been determined by a largely empirical approach. In the early 1960's, however, the idea of scientific engineering was conceived and the techniques of systems engineering were applied to manufacturing for the first time.

A system is a collection of interrelated parts which act together in accordance with a set pattern. The set of procedures employed by the management, the information networks and the equipment available within a company may be regarded as parts of a system responsible for turning raw inputs into material product ouputs. This so-called manufacturing system is concerned with all elements of design, planning, control, machining, assembly and testing processes. It consists of a system in its own right made up of individual sub-systems connected by paths for communication. Information and energy are exchanged or shared between the parts and this in turn implies that the parts change with time and the system is dynamic. This is particularly true of manufacturing systems: changes in economic policy and the rising costs of materials and labour together with the social trends producing shortages of skilled workers have forced changes in the way many companies approach the problems of manufacture. The purpose of a manufacturing system is to convert inputs and outputs. Its performance in achieving this may be evaluated in terms of two criteria; the <u>production efficiency</u> measured in terms of labour and machine productivity, rate of stock turnover, delivery performance, etc., and <u>flexibility</u>, which is a measure of the ability of the system to respond to changing demands and resources. Many companies, in particular long-established concerns operating in areas of rapid technological advance, are finding that their manufacturing systems which evolved from intimate man-machine relationships of the 19th century are no longer efficient in the face of increased competition nor flexible enough to keep pace with changing markets and technology. Machine tools and personnel have been added to increase capacity, ageing equipment has been replaced by more productive plant and new skills have been developed but very rarely have changes been made to the actual manufacturing system laid down when the manufacturing unit was established.

By adopting the systems engineering approach it has been found that manufacturing activities can be represented by <u>mathematical models</u> which permit the use of powerful tools for synthesis, analysis and optimisation. This not only permits a greater understanding of increasingly more complex systems but also permits more rapid changes to existing systems and reduces development times.

10.2.3 <u>Activities and Communications</u>

The activities within an engineering manufacturing system may be divided into five broad areas:

* product identification, specification and design
* production scheduling and forward planning
* production planning and control
* actual manufacturing processes
* inspection and quality control

These areas are interconnected by routes for the flow of products and information. If the system is to function effectively it must have facilities and methods not only for controlling the physical flows but also for the timely generation, collection and communication of information, and many of the problems in manufacturing are caused by poor communications between areas of activity.

The traditional conveyor of both physical items and information is man. It is now held that the degree to which it is possible to optimise the performance of a manufacturing system is directly related to the degree to which communications within the system can operate without human intervention. Planning for optimisation in manufacturing is therefore synonymous with automating the manufacturing system from the design concept to the finished part. The area of manufacturing which has attracted most attention to date is the most labour intensive area, the actual production process, and it is in this area that the most significant advances in automation have been made.

10.2.4 <u>Types of Production</u>

Before considering the extent to which the automation of production processes has developed, it is necessary to identify the different types of production process employed in manufacturing engineering. There are three main types of production, namely, job production, batch production, and flowline production. All three tend to be closely associated and may overlap in many circumstances.

<u>Job production</u> describes the method by which single articles are manufactured. All engineering concerns, whatever their nature, are involved at some time or other in job production, be it the manufacture of small components required for maintenance of plant, the production of prototypes or tools, small jobbing contracts for other concerns or large-scale job-type production such as shipbuilding. The general characteristics of job production systems result from their general-purpose nature and the wide variety of work they must perform. Usually a wide range of general-purpose, versatile machinery and equipment is available together with a staff of highly-skilled personnel and a permanent store of standard materials and components to permit the manufacture of as great a variety of work as possible at short notice. Because of the general nature of the equipment used, however, and the lack of time for detailed optimisation of each job, job production systems tend to be inefficient in terms of manpower and machine productivity. The fluctuating demands on a job production system make it necessary for the system to be highly flexible and change rapidly to suit each particular job. This is usually made possible because individuals, or small teams, are given responsibility for parts, or the whole of the job from beginning to completion and, therefore, the communication problems caused by transfer of authority, information and goods from one section of the system to another can be avoided.

<u>Batch production</u> may be defined as the manufacture of a product in small or large batches or lots, by a series of operations, each operation being carried out on the whole batch before any subsequent operation is started. Batch production is by far the most common method of working in manufacturing industry and it is estimated that approximately seventy-five per cent of all parts produced by the metalworking industries are produced in batches of less than fifty. Batch manufacture is almost universally accomplished by issuing components into manufacture on an 'operation' basis; that is, the work to be done is split down into separate operations involving perhaps five to thirty operations per part. Each of these operations may involve passing the part from one manufacturing process to another and, even if many of the operations are confined to one machine tool, changes in set-up or position of the workpiece on the table are very frequent. These changes in set-up mean that the machine is not cutting during this period and the overall efficiency may be as low as fifteen to twenty per cent. When the workpiece is passed from machine to machine, this situation gets much worse as, even with good organisations, it is rarely possible to manage a large machine shop in such a way that components spend less than one day between operations. To ensure that all components produced in a batch are identical, it is necessary to employ strict inspection procedures which add to the inefficiency of the process. Batch manufacture has several major disadvantages caused by the delays and movements between operations. These communications problems

include:

* large amounts of work in progress develop which involve large capital investments.
* large production storage areas and generous transport facilities are needed and a very effective planning and control system is needed to meet production deadlines.
* comparatively long production periods are needed due to the time that each batch has to wait before proceeding from one operation to the next.

Batch production presents the greatest problems in manufacturing due to the combination of poor efficiency and communications with the need to maintain a high degree of flexibility to enable a continuously altering plan of work output to be applied.

Flowline production is the manufacture of a product by a continuous series of operations, each article going onto a succeeding operation as soon as available. Flowline methods are usually only applied when components are required in very large numbers over long periods of time. The maufacturing system tends to be very rigid, and depends heavily on large financial investments on capital equipment which is designed and arranged, with knowledge of the type of component to be produced, to operate at optimum efficiency.

10.2.5 Mechanisation and Automation

The word automation, which is generally used when referring to increasing the efficiency of a manufacturing system, is a relatively new word and was first coined in 1947 to describe the automatic handling of work pieces. The concept to which it is applied is not new however; in the 19th century technological advances lead to mechanisation of production processes which greatly increased productivity.

The earliest examples of mechanisation in engineering were the use of multi-spindles and power feeds in machine tools. Later developments included sequence-controlled machines which, once set up, could produce large numbers of identical components faster, and at higher rates, than manual machines. These early efforts, however, although greatly increasing productivity and efficiency were generally purely mechanically operated and worked in isolation from one another.

Around 1940, the next step forward was made with the introduction of transfer lines for the machining of aircraft engine components. These consisted essentially of a group of drilling, reaming, tapping and milling machines arranged along both sides of a conveyer along which workpieces, clamped onto special holding fixtures, were moved from work station to work station where different operations were performed. All the operations were interlocked so machining could not occur unless all the workpieces, one at each work station, were positioned correctly and transferring was impossible until all tools and clamps were clear. The transfer line machines thus performed mechanically both the coordination and work handling functions normally carried out by production control and progress personnel.

Although in-line transfer machining systems developed rapidly and proved to be a tremendous step forward in increasing productivity, they did not entirely overcome one of the biggest problems in all machine shops, that of the actual handling of workpieces. Handling between workstations was avoided but someone still had to load components at the first station and unload them at the last and work still had to be moved to and from the transfer line. Furthermore, developments in metal-cutting technology brought about the condition in which it was possible for the machine to turn out work faster than a man could load or unload it. This meant that machines were not used to their maximum capacity and handling time was out of all proportion to machining time.

A second problem with the mechanised transfer line machines was that faults caused by, for example, worn or broken tooling were only detected at some later inspection stage by which time a great deal of scrap may have been produced. The corrective action, replacing the tools and resetting the machine, was performed manually and involved long delays during which the entire line was stopped and no production was possible. The only way to avoid faulty components was to do preventative maintenance on the line and change tools at predetermined regular intervals whether worn or not. Machine tools have now been designed with integral loaders, feeders and unloaders, inspection devices and tool-wear compensation systems. These, combined with improved conveyor systems and industrial robots, have formed the basis of production systems in which it is possible for work to progress from raw stock to finished parts without being touched by hand. Under such conditions it is essential that there is a method of telling whether machines are performing to programme. An automatic system may thus be defined as one which will carry out a pre-set programme or sequence of operations, at the same time measuring and correcting its actual performance in relation to that programme.

By automating work handling, machine tool and information feedback systems, it has been possible to attain very high levels of efficiency in manufacturing. The method of obtaining this efficiency has resulted in very expensive systems, rigidly designed for the production of specific items and only of use in the mass production sectors of industry.

The problems in batch manufacturing still remain and intensive efforts are being made to apply the techniques of automation to all parts of batch manufacturing systems from product design to supply of finished parts.

10.2.6 Numerical Control

The early automatic manufacturing systems were based on special-purpose machines and work handling equipment designed and constructed to carry out a single job with little or no variation allowed, with the specific purpose of obtaining high output of accurately made products. Such systems are of little use in batch manufacture.

One of the most significant attempts at applying the technique of automation into batch manufacture was the introduction in the 1950's of numerically-controlled machine tools. These are a general class of automated machine tools which do not rely on orthodox mechanical means for sequencing and positioning functions but are based on more versatile and sophisticated means for programming which depend heavily on applied electronics. The problem of controlling a machine tool can be divided into the sequencing function, which includes, for example, spindle drive, speed selection, tool selection and the control function, which includes cutting rate and tool position. In a simple sequence-controlled machine, operations may be selected by means of pins inserted into a matrix board, tool positions being fixed by mechanical stops. Such machines, which are usually referred to as 'plugboard auto's', although relatively cheap to buy and simple to control, require a longer time for set-up than more sophisticated numerically-controlled machines in which a control programme, usually in the form of a punched paper-tape, provides the sequencing and positioning commands and servo systems provide the necessary controls. As a result, plugboard-controlled machines tend to be used for batches of more than one hundred and tape-controlled NC machines are most suitable for smaller batches.

The advantages of the NC machine over conventional equipment in batch manufacture include:

* the ability to produce components of consistent geometry and quality at high rates for long periods so reducing scrap and rework. It is possible that one NC may replace three or four conventional machines so reducing the cost of floorspace and manpower.
* the use of long control programmes and automatic tool changers make it possible to combine many conventional operations into one NC operation and so reduce the amount of work handling and ultimately the lead time of a component and the amount of work in progress.
* rather than relying on jigs and fixtures for geometrical information, as do conventional machines, the NC machine obtains all the required information from the control programme. As the cost of modifying the control programme is considerably less than that of modifying jigs etc., NC machines make design changes more easy and cheaper to embody and so increase the flexibility of the manufacturing system.

As NC machines became more sophisticated it was necessary to employ aids to their programming. This led to one of the first applications of computers in manufacturing.

10.2.7 Computer-Aided Manufacture

The developments in computer technology during the early 1960's lead to a rapid expansion in the number of applications of computers. Originally it was thought that they would only be used in scientific applications but, as they became more easily used and more readily available, many other applications were found including many in business and industrial manufacturing. About ten years ago, the use of computers had advanced so far that the term Computer-Aided Manufacture (CAM) was coined to describe the application of computers in manufacturing systems. The mass production industries, such as those involved in petrochemical refining, were able to make use of the capacity of the computer for large-scale data storage and rapid calculation and soon developed computer-based manufacturing systems. More recently, there have been many applications of computers in batch manufacturing. These may be divided into the off-line processing of data pertinent to produce design and manufacturing planning, and the on-line control of production processes and information.

Off-line applications are those in which the computer is remote from the manufacturing system and operates independently of it. Initially, computers found their greatest uses in manufacturing in finance and accounting and in the preparation of NC part-programmes, since these areas have customarily used computational methods and calculating equipment. Later developments have included computer-based systems for production planning and control and for computer-aided design of components.

It is convenient when considering CAM systems to consider planning and control as separate sub-systems. The computer-based production planning systems that have been developed typically include programme modules and procedures for:

* requirements planning, which will calculate the required capacity and materials from a sales forecast or order load.
* capacity planning which will roughly calculate a schedule for loading machine tools and other resources with a planning horizon of the order of one year and a planning interval of the order of one month.
* scheduling to provide detailed loading of each resource group in a short period with a planning horizon of the order of one week and a planning interval of the order of a few hours.

Typical computer-aided production control systems involve modules and procedures for:

* purchase control including replenishment of raw materials, stock control and purchase order control.
* production and assembly control including materials control, load control, inventory control, tool control, job control and dispatching.

The off-line applications are usually performed by what is known as batch processing on a large centralised computer. Data are collected from the plant manually and, together with new production orders and work loads, these serve as input to the planning system which may be run daily, weekly or monthly depending on the production type. The output from the computer generally consists of listings from which the detailed schedule and necessary orders can be extracted. The off-line type of applications are usually the result of applying the computer to the manual procedure. The speed and accuracy of the computer permit the optimisation and simulation of plans which would be impossible in the time available using manual means. The discipline of installing a computer system also results in improvements which are due to the fact that procedures and activities are analysed and improved when converting manual procedures into computer logic.

In the early 1960's computer-based systems were developed for the direct digital control of continuous fluid processes. These were examples of on-line applications of computers in manufacturing in which the computer is an integral part of the manufacturing system. On-line systems may be divided into two sub-systems: those for monitoring and information systems and those for control of manufacturing processes. The purpose of monitoring and information systems is to register and report production data. This may be data concerning active, idle and breakdown times of different machines, inventory transactions or job status and is collected automatically via direct connections to the computer. These systems provide management with up-to-date information regarding the status of the manufacturing resources and so increase the flexibility of the manufacturing system by easing the decision-making processes. There have been many developments in the computer-based control systems which include:

* the sequence-control of a production line, which may involve the knowledge of production data including the number of pieces produced, cycle time, and idle time.
* Computer Numerical Control (NC) of machines which is numerical control in which the hardwired conventional control is replaced by a minicomputer programmed to perform the control functions. The power and versatility of the minicomputer and its inherent reliability permits more sophisticated control at a lower cost than is possible using conventional transistor-based logic.
* Adaptive Control (AC) of machine tools in which the computer is used to measure, for example, cutting forces and speeds and to control the axis motion and spindle speed accordingly so as to maintain the optimum metal removal rate.
* Direct Numerical Control (DNC) of machine tools which is the connection of several NC machines to a central digital computer for part-programme distribution and storage.

The characteristic feature of on-line systems is that a dedicated computer is used in real-time mode; that is the computer is available at any time on demand to perform its function.

10.2.8 Integrated Computer-Aided Manufacturing Systems

Computers are being used in ever increasing numbers in manufacturing systems but in a somewhat disjointed fashion. New applications have solved particular problems but the overall contribution of the computer has often been less than forecasted. It has also become apparent that it is not enough to superimpose computer technology and techniques onto traditional manufacturing systems. The most promising concept for solving the problems of efficiency and flexibility is the Integrated Computer-Aided Manufacturing (ICAM) system. Such a system would be based on work stations, interfaced by automatic handling systems, which have been designed from the 'floor up' to efficiently interface with the digital computer. All aspects of the manufacturing activity including detailed design, specification, manufacturing engineering, materials management, production of parts, assembly, test, warehousing, sales and service, would be controlled by individual modules of computer software, all of which would be linked together in a hierarchical system. ICAM is a total technology which will involve tremendous amounts of software and will be evolved over a fairly long period of time. The development will be evolutionary: each module of software will be self-justifying and will perform a useful role within an existing manufacturing system.

10.3 FUNCTIONAL AND INFORMATION MODELS OF A TYPICAL BATCH MANUFACTURING SYSTEM

10.3.1 Structured Analysis

The models presented in this section are based on the structured analysis design technique, SADT, of SofTech. This technique provides a powerful tool for describing the relationship between activities and data in a system. There are two fundamental models used in SADT: firstly, the <u>functional model</u> which is concerned principally with activities and, secondly, the <u>information model</u> which is concerned principally with data.

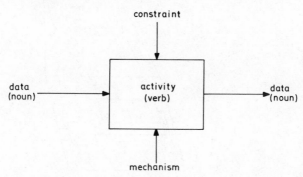

Fig. 10.1 Basic structure for functional (activity) model

The typical building block in a functional model is shown in Fig. 10.1 in which the <u>activity</u> (a verb) is contained within a box which receives <u>input data</u> (a noun) and produces <u>output data</u> (a noun). The top entry to the box is a <u>control</u> or <u>constraint</u>, whilst the bottom entry to the box is a <u>mechanism</u>.

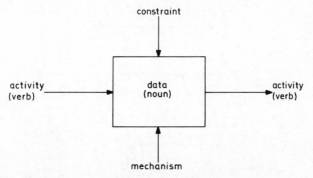

Fig. 10.2 Basic structure for information (data) model

On the other hand, the typical building block in a data model is shown in Fig. 10.2 in which the <u>data</u> (a noun) is contained within a box which receives <u>input activity</u> (a verb): these data are then used or consumed by an activity on the right-hand side of the box. As in the previous case, the data in the box can be subjected to both constraints and mechanisms. Functional models and information models are designed to have not less than three boxes and not more than six boxes, and the output of any box can become an input, constraint, or mechanism to any other box or model. The modelling of a complete system is accomplished by a <u>decomposition</u> of an uppermost model called AO in the case of a functional model and DO in the case of a data model. If, for example, AO comprised four activity boxes, A1 to A4, then each of these boxes with their associated inputs, outputs, constraints and mechanisms could themselves be expanded to the next level of decomposition.

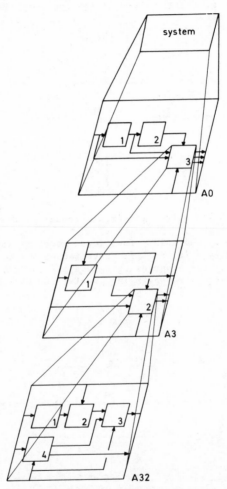

Fig. 10.3 System decomposition

Thus, if activity A2 comprised five identifiable activities then these would be called A21, A22, ..., A25. This process of decomposition can be carried on to more and more detailed levels as shown in Fig. 10.3, subject only to the constraint that inputs, ouputs, constraints and mechanisms to any box must be maintained at the next level of decomposition: it is possible in some cases to introduce local operations on boxes and these are usually denoted by enclosing them in parentheses. In association with a structured decomposition, it is normal to produce a node list which is a list of names given to each of the activity boxes and/or data boxes. There are many symbolic notations and drawing procedures which can be used to highlight detail or to simplify the presentation of a model.

Two of the more common simplifications are shown in Figs. 10.4 and 10.5. Thus, in Fig. 10.4(a), the output of box 1 is an input to box 2, and the ouput of box 2 is an input to box 1: the mutual exchange of data can be simplified as shown in Fig. 10.4(b). Similarly, in Fig. 10.5(a), the output of box 1 is a constraint to box 2, and the output of box 2 is a constraint to box 1: this output/constraint relationship can be redrawn as shown in Fig. 10.5(b).

A rigorous checking of both the functional and the information models is essential, as they contain complementary descriptions of the same system and must be completely compatible. It should be noted that the models are essentially qualitative and that time is not shown explicitly on the diagrams.

For a full rigorous description of this modelling technique reference should be made to appropriate literature [1,2,3,4,5] and to SofTech Inc.

10.3.2 Illustrative Example

The selected illustrative example is based on a full structured decomposition of a company engaged in batch production of sheet-metal components for the consumer market. The production facilities of this company are organised on a cellular basis rather than on a traditional functional layout. Whilst the company was modelled from the uppermost levels, AO and DO, down to the sixth level of decomposition, diagrams for only the first three levels are included as shown in the node lists presented in Tables 1 and 2: only those decompositions indicated by asterisks are presented in this chapter. Only the part of the model appropriate to the control of a job shop concerned with the batch manufacture of sheet-metal components is described. This area of the operation is shown as functional models in Figs. 10.6 to 10.9 and as information models in Figs. 10.10 to 10.13. The brief notes accompanying these models are as follows:

AO Run Company (Fig. 10.6)

This diagram overviews the operation of the company used for the illustrative example. The finance shown as an input to the three activities, A1, A2, and A3, represents borrowing and share capital in the form of additional investment in the company. The formulate strategy activity, A1, sets the broad objectives and operational framework for the company against the expectations of the market place and the community. This strategy forms the basis for the formulate company plan activity, A2, constructed to meet the company requirements in terms of financial performance. The implement company plan activity, A3, is the operation of the company to meet the planned objectives in terms of satisfactions, profits and payments. All these activities are the concern of the company's senior management, as indicated by mechanisms.

A3 Implement Company Plan (Fig. 10.7)

Implementation of the company plan is split into four activities. The financial activities operate as a service and control function for the prime activities of design, production and marketing. The payments leaving 'Manage Finance' are wages, payments to suppliers, dividends and taxes. The company has a standard product range and manufactures no bespoke products. Customer demand is reflected in the design of new products and the redesign or modification of existing products.

A33 Produce Products (Fig. 10.8)

The production of products is broken down into five activities. The preparation of the production programme is performed by senior management who produce a twelve-monthly programme and the associated production policy. There is a feedback of performance between this activity and the Plan and Control Production activity. The spares output from manufacture can be single components and sub-assemblies. A standard computer package is used as an aid to production planning and control.

A332 Plan and Control Production (Fig. 10.9)

The company's computer production control package explodes the product and compares it against the production programme to determine gross requirements. Control of stock is done manually and the data are fed into the computer system for determination of the net requirements. Every four weeks the computer produces a six-weekly forecast of net requirements which is broken down, by a scheduler, into three two-weekly workloads giving details of parts to be made and batch quantities for each machine group or cell. Each two-weekly workload is then scheduled in detail for each cell using conventional bar charts.

DO Run Company (Fig. 10.10)

This diagram overviews the information and facilities required to run the company. Based upon the market conditions, the available capital and resources, a corporate strategy is developed. The company objectives are satisfied through the preparation of company plans and policies which are used to drive the corporate organisation for the achievement of satisfaction, payments and profits.

D3 Company organisation data (Fig. 10.11)

In-line with company policy, the company accounts are used as the basis for preparation of budgets for design, production and marketing. The interaction shown between boxes 3 and 4 indicates product production with feedback in the form of sales forecast preparation in response to market needs. The interaction between boxes 2 and 3 indicates product design with feedback on product producability. The interaction between boxes 2 and 4 indicates product design with feedback on the market suitability of product designs.

D33 Factory organisation and resources (Fig. 10.12)

The design of products and the requirements for these products are the basis of the company organisation. The twelve-month production programme is the input to the scheduling control system which is aided by a computer package for the determination of procurement and manufactured in-house items.

D332 Scheduling and control system (Fig. 10.13)

The diagram shows information required to plan and control production in the sheet-metal shop. The net requirements are determined monthly by the computer package based on stock levels and production requirements. A six-weekly work load is also produced using the computer. The preparation of weekly and daily schedules for the cells is carried out manually.

These models have been decomposed so that employees at all levels in the company can fully understand and appreciate the operations and procedures in which they are intimately involved. In addition, the model can be used as a basis for simplifying and rationalising the company organisation and the control procedures used in the manufacturing process. In particular, this model highlights those areas where the company computer is used only in isolation and that considerable scope exists for the integration of the overall operations of the company using on-line real-time computers.

10.4 ACKNOWLEDGEMENTS

The author would like to acknowledge the contribution made to the preparation of section 2 by Dr Duncan McCartney, postgraduate of the University of Salford. In addition, the work of Mr Ken Swift and his colleagues in the Industrial Centre Ltd at the University of Salford formed an invaluable basis for the models of the batch manufacturing system presented in section 3.

10.5 REFERENCES

1. ROSS, D.T., GOODENOUGH, J.B., IRVINE, C.A.: 'Software Engineering: Processing, Principles, and Goals', Computer, May 1975, pp.17-27.
2. IRVINE, C.A., BRACKETT, J.W.: 'Automated Software Engineering Through Structured Data Management', I.E.E.E. Transactions on Software Engineering, January 1977, v.SE-3, No.1.
3. ROSS, D.T.: 'Structured Analysis (SA): A Language for Communicating Ideas', I.E.E.E. Transactions on Software Engineering, January 1977, v.SE-3, No.1.
4. ROSS, D.T., SCHOMAN, K.E.: 'Structured Analysis for Requirements Definition', I.E.E.E. Transactions on Software Engineering, January 1977, v.SE-3, No.1.
5. ROSS, D.T., BRACKETT, J.W.: 'An Approach to Structured Analysis', Computer Decisions, September 1976, v.8, No.9.

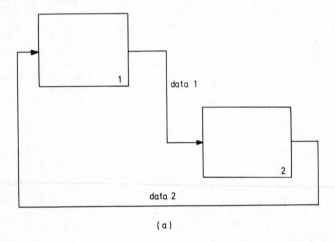

(a)

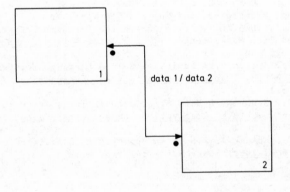

(b)

Fig. 10.4 Mutual exchange of data

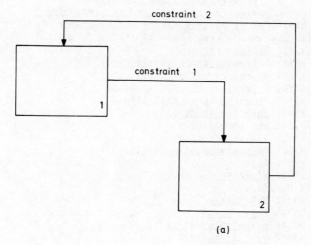

(a)

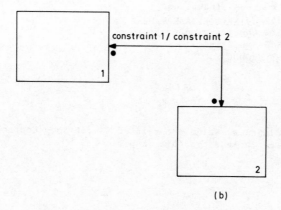

(b)

Fig. 10.5 Mutual exchange of constraints

TABLE 10.1.   ACTIVITY NODE LIST

AO RUN COMPANY*
    A1 FORMULATE STRATEGY
    A2 FORMULATE COMPANY PLAN
    A3 IMPLEMENT COMPANY PLAN*
        A31 MANAGE FINANCE
        A32 DESIGN PRODUCTS AND TOOLS
        A33 PRODUCE PRODUCTS*
            A331 Prepare production programme
                A3311 Plan production methods, times and costs
                A3312 Evaluate batch frequency and quantity
                A3313 Assess resources and operational
                        characteristics related to company plan
                A3314 Set production policy and programme
            A332 Plan and control production*
                A3321 Explode product
                A3322 Determine gross requirements
                A3323 Assess inventory levels
                A3324 Determine nett requirements
                A3325 Prepare period work loads
                A3326 Schedule in detail
            A333 Procure commodities
                A3331 Vendor selection
                A3332 Determine order quantity
                A3333 Buy
                A3334 Expedite
            A334 Manufacture
                A3341 Make parts
                A3342 Stock parts
                A3343 Assemble products
                A3344 Test products
                A3345 Stock products
            A335 Progress production
                A3351 Collect progress data
                A3352 Compare plan/progress
                A3353 Expedite
        A34 MARKET PRODUCTS

* Only these diagrams, which are related to Job Shop Control in Batch
Manufacture, are included in this chapter.

TABLE 10.2.   DATA NODE LIST

DO RUN COMPANY*
  D1 MARKET POSITION, CAPITAL AND RESOURCES
  D2 COMPANY OBJECTIVES
  D3 COMPANY ORGANISATION DATA*
    D31 <u>COMPANY ACCOUNTS</u>
    D32 <u>PRODUCT CONCEPTS AND REQUIREMENTS DATA</u>
    D33 <u>FACTORY ORGANISATION AND RESOURCES</u>*
      D331 Production requirements and policy
        D3311 <u>Component and product designs</u>
        D3312 <u>Annual usage values</u>
        D3313 <u>Available company resources and</u>
              <u>organisation</u>
        D3314 <u>Arrangement of production requirements</u>
      D332 Scheduling and control system data*
        D3321 <u>Product drawing data and parts lists</u>
        D3322 <u>Bill-of-materials and production</u>
              <u>programmes</u>
        D3323 <u>Part stock and work-in-progress</u>
        D3324 <u>Gross requirements and stock list</u>
        D3325 <u>Six-weekly work loads</u>
        D3326 <u>Two-weekly work loads</u>
      D333 Bought-out items
        D3331 <u>Vendor information, price, delivery</u>
        D3332 <u>Order quantity economics</u>
        D3333 <u>Orders to vendors</u>
        D3334 <u>Delivery performance</u>
      D334 Process and job-shop data
        D3341 <u>Sheet-metal shop</u>
        D3342 <u>Stores</u>
        D3343 <u>Assembly shop</u>
        D3344 <u>Product specification</u>
        D3345 <u>Product stores</u>
      D335 Throughput performance
        D3351 <u>Work-in-progress</u>
        D3352 <u>Job status</u>
        D3353 <u>List priorities</u>
    D34 <u>PRODUCTS</u>

* Only these diagrams, which are related to Job Shop Control in Batch
Manufacture, are included in this chapter.

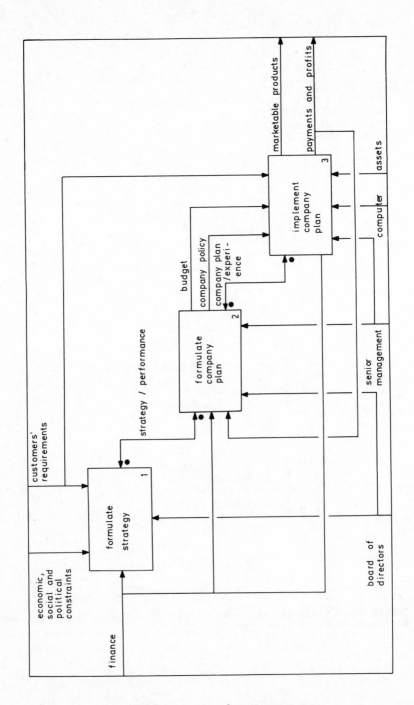

Fig. 10.6 Activity of run company

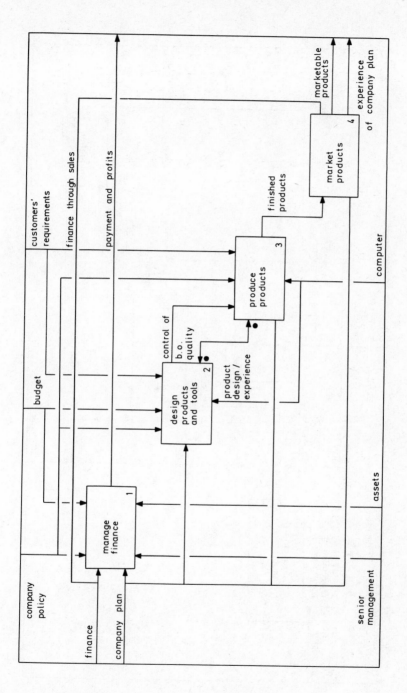

Fig. 10.7 Activity A3, implement company plan

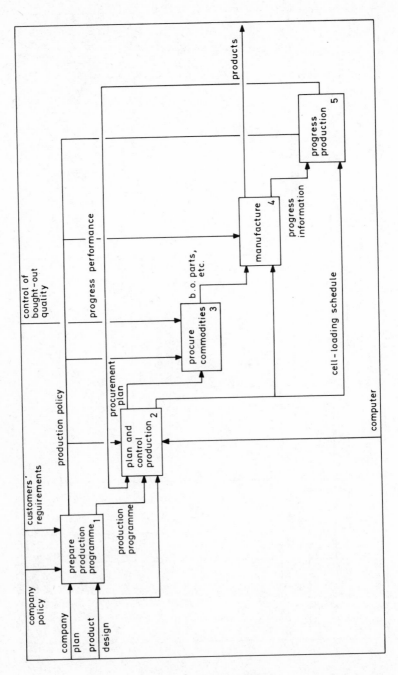

Fig. 10.8 Activity A33, produce products

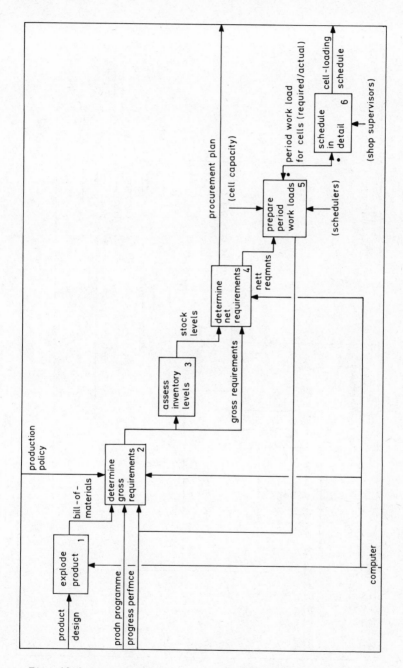

Fig. 10.9 Activity A332, plan and control production

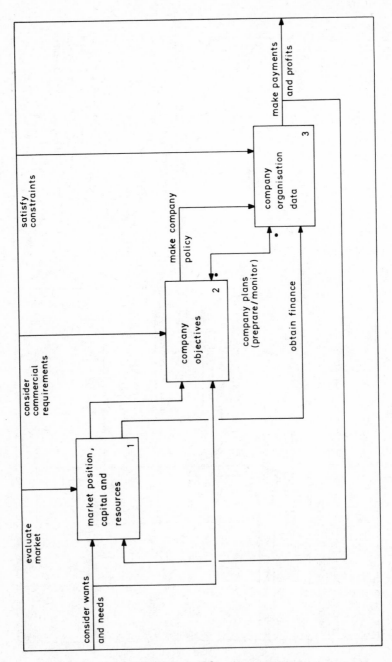

Fig.10.10 Data DO, run company

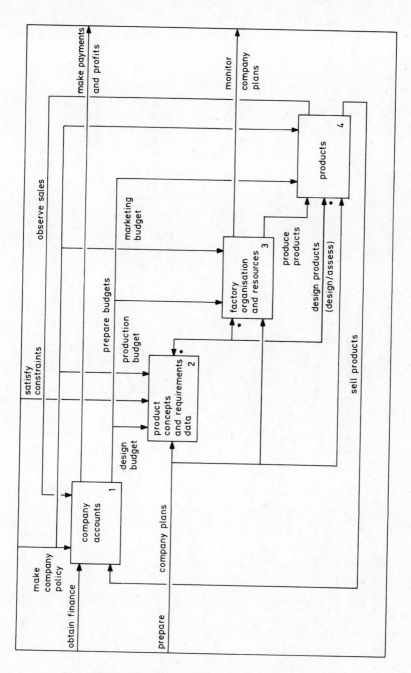

Fig. 10.11 Data D3, company organisation data

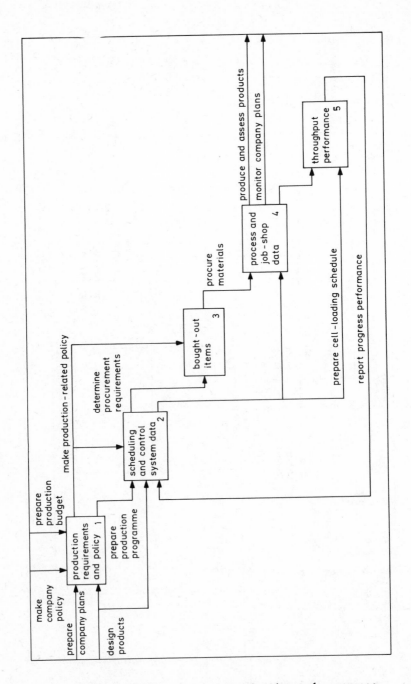

Fig. 10.12 Data D33, factory organisation and resources

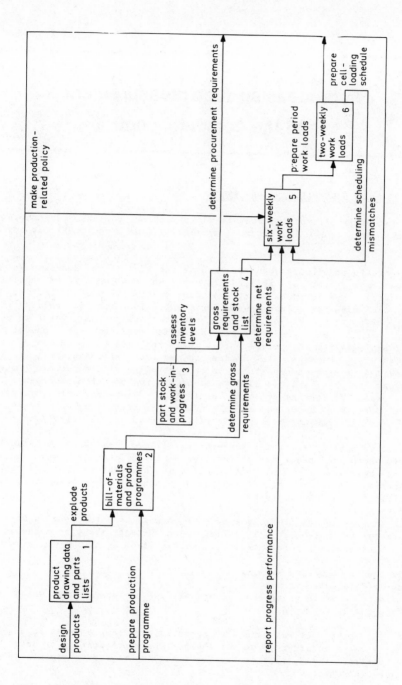

Fig. 10.13 Data D332, scheduling and control system data

Chapter 11

In process surface measurement
and the computer control

11.1 WHY CONTROL? PAST AND PRESENT

In the early part of the industrial revolution, engines, instruments, weapons and such like were individually made – equipment was custom built.

This state of affairs changed at about the time of the American War of Independence. Lack of skilled armourers who were capable of repairing individually made weapons meant that many beautiful pieces had to be discarded with dire consequences for the British. The USA realised this short-coming very early on, so they began to design and make weapons which could be easily repaired. This lead to the concept of interchangeability of parts. This was the start of metrology in manufacture, because in order to ensure interchangeability the dimensions of each component had to be controlled within certain limits. Previous to this, individual dimensions had not been too important. The whole weapon or instrument could be fitted: dimensions of industrial parts could be adjusted to take account of extra long or extra short bits. With interchangeability this fitting could not be allowed.

Since this early start in engineering, there has been a dramatic growth both in the types of function demanded from engineering parts, and in their severity. This has resulted in the need to control other geometric features of a component such as the roundness, the straightness and the texture.

In the last decade there has been a change in the philosophy of manufacturing metrology. This has been brought about by a number of extra constraints to those mentioned above. These are:

* the need to conserve energy;
* the need to conserve materials;
* the need to avoid using skilled operations on components which are subsequently scrapped.

In the past, it was generally acceptable to sample a few components out of a batch in order to decide whether the manufacturing process was under control. Today, this is far too risky and expensive especially for critical components.

The result of these constraints has been a shift towards one hundred percent inspection of the part while it is being made, and thereby applying in-process correction - the difficulty is what to measure and how to measure it.

Another aspect of manufacturing metrology which has emerged with growing importance in the past few years is that of the control of positioning. This has increased in importance since the advent of the concept of automatic assembly. The problem facing the metrologist in this area is whether or not closed-loop control rather than open-loop can be achieved economically and effectively.

In the following sections some modern approaches and the reasons for them will be briefly explained.

11.2 CONTROL WHAT?

Apart from the need to measure dimensions and positions directly in order to guarantee assembly, there are a number of new geometrical parameters which are now being measured with the express purpose of improving the economics of manufacture. The economics of machinig is mainly concerned with tool wear and machine down-time. It is now becoming possible to control these parameters using the component geometry - and the surface deviations in particular.

Consider the case of tool wear. A classical way of determining tool degradation has been to measure the forces on the tool. One major disadvantage of this has been the fact that the force transducer is invariably quite remote from the tool tip. An ideal situation would be to keep a continuous watch over the tool tip geometry during cutting. Because of debris, chips, and coolant this is impossible. The next best thing to do is to examine the surface produced by the tool. This surface, which is in the immediate vicinity of the tool, will be a replica of the tool tip itself and is much more accessible for investigation. How does this help to assess tool wear?

Assume that the tool feed mark can be represented by $f(x)$ in the absence of other influences like machine tool vibration, it is true to make the observation that $f(x) = f(x + n\lambda)$ where n is an integer and λ is the tool feed distance. The value of n will vary from 1 to cover the whole of the machined surface. A wear scar on the tool will, like the general tool geometry, repeat itself every tool feed spacing λ. One tool scar at a distance of α_1 from the arbitrary axis can be represented by $k_1(x+\alpha_1)$ where k_1 is an impulsive type of signal being relatively short in spatial duration. The waveform replicated in the component becomes modified to $f(x) + k_1(x+\alpha_1) \equiv f(x+n\lambda) + k_1(x+\alpha_1+n\lambda)$. For s scars on the tool the replicated waveform is built up to become

$$f(x) + \sum_{i=1}^{s} k_i(x + \alpha_i). \tag{11.1}$$

If the Fourier transform of $f(x)$ is compared with that of

$$f(x) + \sum_{i=1}^{s} k_i(x + \alpha_i)$$

then one interesting and useful fact emerges.

The ratio of the sizes of the harmonics compared with the fundamental (equivalent to the reciprocal of the feed) is different. In fact the harmonics become progressively more important as 's' increases: in other words as the wear increases (see Fig. 11.1).

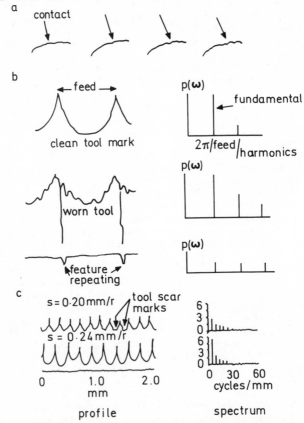

Fig. 11.1 Use of random process analysis to reveal tool wear

This is because each wear scar being repetitive and pulse-like can be considered to be an impulse train of period λ. This impulse train has for its Fourier transform an impulse train in spatial frequency in which the harmonics are of exactly the same height as the fundamental. This is true for all the scars. The ratio of harmonics to fundamental for $f(x)$ however is not the same. This depends solely on the shape of the tool and the harmonics are very small. If the ratio of harmonic amplitudes to the amplitudes of the fundamental is called W then this ratio W increases as the number of wear scars increases because the relative height of the fundamental to the harmonics for these impulsive wear scar trains approaches unity unlike that of the clean-cut tool impression.

It is not only the harmonics which can be important in machine performance, the sub-harmonics are also important. This time, however, it is not the tool-wear which causes changes in their relative amplitude, as it is for the harmonics, it is the machine itself which influences the amplitudes. Consider first the axial sub-harmonics —

these are affected very strongly by the slideways of the machine   tool.
Incipient failure of the headstock bearing of a lathe, for example, can
show itself as the emergence of a sub-harmonic of the tool feed spacing
long before there is any noticeable degradation in performance [1]
(Fig. 11.2).

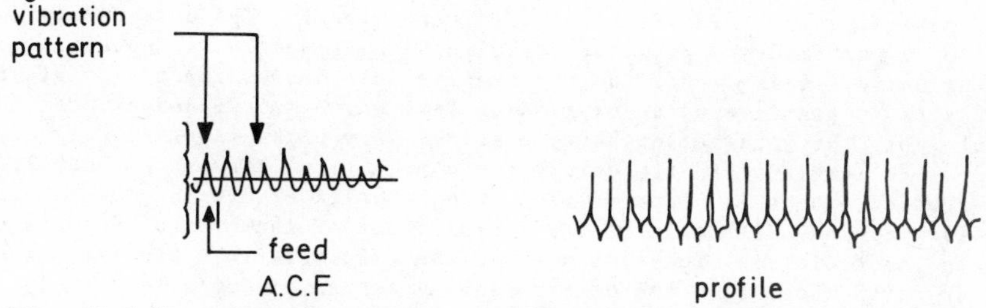

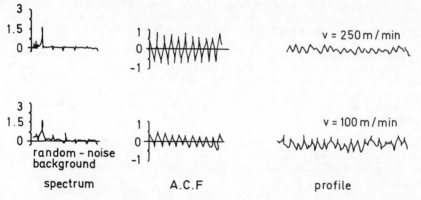

Fig. 11.2 Use of random process analysis to reveal machine vibration
          due to chatter of bearings

Lobing in the circumferential geometry also reflects not   only   the   way
the component is being clamped, but also rigidity of the machine frame.
Highly elastic structures can often produce chatter   marks   around   the
component.

   Again, examination of the spectrum can often indicate the suitability
of the process. Metal tearing, which can be produced because of a   slow
cutting speed or unsuitable metal property specification, reveals itself
as a 'white noise' component in the spectrum: the   effective   base   line
gets raised (Fig. 11.3)

Fig. 11.3 Use of random process analysis to reveal surface condition due
          to built-up-edge

Sometimes even in the cutting process this random element can be
beneficial rather than detrimental to performance. This is so in the
case of diamond turning - one of the fastest growing manufacturing
processes, where getting fine turning is often better achieved by
allowing the back end of the tool to burnish out the tool mark
impression in its passage across the component. This shows itself as a
destruction of the periodic component of the spectrum and an increase in
the random. A measurement of the random element can, therefore, be used
in principle to control the amount of burnishing. This is not a simple

matter even then because this involves controlling the cutting tool angle relative to the workpiece to within a degree.

The random element can also be important in other manufacturing processes, such as grinding or bead blasting. The Fourier transform of the power spectral density shows this very clearly. The autocorrelation $A(\tau)$ of a part made by grinding reveals the unit event of manufacture. In grinding, this would be the average grit impression $u(\tau)$. This is modulated in practice by an expression $T(\tau)$ which is dependent on the nature of the interaction between grits. Very often in practical cases this is Poissonian, and the envelope becomes very nearly exponential. Divergences from this shape reflect wheel unbalance, dressing errors and similar machine malfunction. As a general rule of thumb, the region of the autocorrelation function near to the origin reflect process errors whilst further out from the origin machine errors predominate.

The general conclusion which can be reached from this outline is that the autocorrelation function and power spectral destiny are being effectively used as tools for assessing process and machine tool error. The real problem remains, as always, in finding simple economic ways of measuring these statistical functions in the extremely hostile manufacturing environment and thereby be in a situation to control the manufacture.

One other aspect of manufacture which is growing in importance is that of automatic assembly and associated problems of high precision positioning. Another is the use of robots to help in the job of one hundred percent post-process inspection without human intervention. Both of these growth areas have as one of their concepts the use of robots as dynamic jigs or dynamic fixtures – a somewhat contradictory idea. Both rely on the application of robots having a high precision, wide range and yet a quick response time. The fundamental requirement in metrology is that of measuring absolute or in some cases relative position. This is straight forward in large scale applications where component tolerances can be considered to be independent of texture. But in micro-engineering where they are not, the problems are magnified. It is only with the advent of miniaturisation that this manufacturing and assembly problem has arisen.

Assembly and function now depend more than ever before on errors in general shape as well as those of size. Typical parameters are errors in cylindricity, sphericity and all similar 3D uncertainty.

These parameters are becoming more difficult to control during manufacture because of the increased use of multi-purpose CNC machine tools which are not specifically designed to produce individual parts. The larger the CNC machine the more difficult is the control because of the Abbé offset errors introduced in the large mechanical loop.

11.3 SOME MODERN METHODS OF CONTROL

Conventional multigauging methods and stylus technique have been well reported [2]. However, these usually apply to post-process inspection or simply in-situ inspection. The distinction is that although the part may still be in the machine there is no actual machining taking place as would be the case for in-process measurement.

Perhaps the nearest that has been achieved for in-process measurement of texture using stylus methods is due to Dutschke [3]. In this a roller runs against the part to be measured as it is being machined. Inside this roller is a transducer connected to a stylus which protrudes through the roller wall and which contacts the workpiece once per revolution. After a few contacts the output from the transducer gives some idea of the surface finish. Unfortunately, the technique does not give a comprehensive value of texture which takes into account spacings of asperities. It also measures the component circumferentially with the result that roundness errors get mixed in with the surface texture signal.

The measurement of out-of-roundness using stylus methods is well known. However, the technique requires an accurate datum, usually a very good spindle, which is capable of generating very nearly a perfect circle in space. Because of this requirement it is impossible to implement such a method for in-process application because there is absolutely no guarantee that the rotation of a part in a machine will be truly circular. The nearest in-process yet devised but not fully proven in practice makes use of an ingenious new concept [4]. Basically instead of one stylus three styli are used. They are coplanar and held in a yoke in such a way that their line of action has a common point. If the angles between the outer two relative to the middle one are α and β respectively and the sensitivities of the transducers are $a, -1$ and b respectively then the values of these instrument parameters can be chosen in such a way that the sum of the transducer outputs will not see a rotating circle. In other words if this device straddles a perfectly circular part which is being imperfectly rotated the output will be zero. This corresponds with removing the first harmonic component of the signal from the part. This is accomplished if

$$a \cos\alpha + b \cos\beta - 1 = 0$$
$$b \sin\beta - a \sin\alpha = 0$$

(11.2)

However, in achieving this insensitivity to eccentricity errors the out-of-roundness signal has become somwhat scrambled. In fact if the true out-of-roundness is $f(\theta)$ the device sees $s(\theta)$ where $s(\theta) = -f(\theta) + af(\theta + \beta) + bf(\theta + \beta)$ all at once.

It can be shown [4] that if a harmonic analysis is made of the signal emerging from the device $s(\theta)$ such that

$$s(\theta) = \sum c_n' e^{+(n\theta + \psi_n)}$$

and this is compared with a similar harmonic analysis of $f(\theta)$ such that

$$f(\theta) = \sum c_n e^{(n\theta + \psi)}$$

then in order to convert c_n' to c_n and ψ_n' to ψ the c_n' amplitudes are modified by a factor k

where $k = [(a\cos\alpha + b\cos n\beta - 1)^2 + (b\sin n\beta - a\sin n\alpha)^2]^{\frac{1}{2}}$

and the phases ψ_n' modified by a factor γ

$$\text{where } \gamma = \left[\frac{b\text{Sin } n\beta - a\text{Sin } n\alpha}{a\text{Cos } n\alpha + b\text{Cos } n\beta - 1} \right]$$

So the true roundness error of the component can be evaluated in-process despite imperfect rotation. Similar techniques can now be used for all sorts of form error [4]. Needless to say, it has only been in the past few years that such techniques have become possible. Only now is it feasible to carry out this harmonic analysis sufficiently fast to measure the form errors in real time. The same argument applies to other similar calculations which have to be carried out in straightness and flatness error measurements.

The establishment of the features of geometry which can be used to control the process poses the question of how they can be measured. Two distinct problems arise, how to get the information off the workpiece and how to asses it. Such a measurement has the constraints that the workpiece should not be damaged during, or even by, the measurement and it should be fast enough to allow in-process control of the machine.

Traditional techniques would involve the use of contacting stylus transducers to pickup the information from the workpiece and analogue filtering and metering hardware to asses it. In the past this approach has been acceptable because of the relatively crude understanding of what could be achieved, but during the last ten years the need for flexibility in parameters and the need for the ability to control machines has led inevitably to a swing towards digital methods.

From what has been said about the useful random process parameters, the autocorrelation function and the power spectral density, it might be assumed that these were the only parameters worthy of mention. This is not so. These functions of a waveform are spatially orientated and in particular have been derived for use in time-series analysis typical of problems in communication theory. Problems associated with engineering components are more difficult because the part itself is three-dimensional and it invariably has to fit or mate with one or more other similar (or dissimilar) parts in order to perform its intended function – as a bearing or seal or whatever. In these functions the purely longitudinal characteristics are relatively unimportant when compared with the characteristics involved in contact between parts and their relative movement. The geometrical parameters of interest are more hybrid in nature. For example, in contact situations the curvature of the asperities of the surface are of vital importance together with the distribution of asperity heights over the whole surface. In lubrication it is the probability of an asperity occurring at a given height, and hence being in a position to rupture the fluid film, which is of prime importance. Similar difficult parameters are important in other functional situations. Ideally, therefore, what is needed is not simply a means of control over manufacture: this is simply begging the ultimate question of whether the part, however manufactured, will work when put to use.

It is with this realisation in mind that much effort is being expended to make the most use of existing knowledge of digital techniques, random process analysis and tribology theory to devise methods of assesment which will be of most practical use not only to the manufacturer but also to the user. Hence, techniques are being developed which although initially derive correlation or spectral values − in themselves useful in control − also enable a measure of functional prediction to be achieved at the same time. Take, as an example, the measurement of average peak curvature.

This can be shown to be given by

$$C = \frac{3 - 4\rho_1 + \rho_2}{2 \, N \, h^2 \, [\pi(1-\rho_1)]^{\frac{1}{2}}} \tag{11.3}$$

where

$$N = \frac{1}{\pi} \arc \tan \left[\left(\frac{3 - 4\rho_1 + \rho_2}{1 - \rho_2} \right)^{\frac{1}{2}} \right],$$

ρ_1, is the correlation between adjacent ordinates; ρ_2 between alternate ordinates; h is the spacing between digital samples, and the surface roughness (measured as the rms value) is taken to be unity.

Even this parameter, extracted from the correlation function, is an approximation, because it only relates to an isotropic surface. Most real surfaces are neither anisotropic nor isotropic so that there may be more than one correlation. In the general case of anisotropic surface the expected curvature of a two-dimensional summit is given by [5].

$$C = \frac{(1 + a + 2b)}{h^2} \, E \, [z_0/T_s] \tag{11.4}$$

where h is the sample interval

$$a = \tfrac{1}{2}(1 - 2\rho_1 + \rho_2)/(1 - \rho_1)$$

$$b = \tfrac{1}{2}(1 - 2\rho_1 + \rho_3)/(1 - \rho_1)$$

and

$$E \, (Z_0/T_s) = 2\sqrt{[(1 -\rho_1)/\pi]}\emptyset^{(3)}(0;B_4)/\emptyset^4(0;V_4)$$

where

$$B_4 = \begin{bmatrix} 1 - b^2 & a - b^2 & b(1 - a) \\ a - b^2 & 1 - b^2 & b(1 - a) \\ b(1-a) & b(1-a) & 1 - a^2 \end{bmatrix}$$

and

$$\emptyset^{(3)}(0;B_4) = \tfrac{1}{2} - (4\pi)^{-1} \, \text{Cos}^{-1}((a - b^2)/(1 - b^2))$$

$$-(2\pi)^{-1} \, \text{Cos}^{-1}(b(1 - a)[(1 - a^2)(1 - b^2)]^{-\frac{1}{2}})$$

and $\quad \emptyset^{(4)}|a,b| = \emptyset^{(4)}|a,o| + \int_o^b \frac{\delta}{\delta t} \, \emptyset^{(4)}[a,t]dt$

$$= [\tfrac{1}{4} + (2\pi)^{-1} \, \text{Sin}^{-1}(a)]^2 + (2\pi)^{-1} \, \text{Sin}^{-1}(b)$$

$$= \pi^{-2}\int_o^b (1-t^2)^{-\frac{1}{2}} \, \text{Sin}^{-1}(t(1-a)(1+a-2t^2)^{-1})dt$$

This complexity is typical of all parameters of functional significance in engineering and yet it can be evaluated digitally but not by analogue means, and furthermore the values of correlation ρ_1, ρ_2, etc., can be utilised for control. The only alternative means of obtaining parameters of functional significance, such as the summit curvature, would be to devise schemes of taking multiple tracks across the surface while the part is being made and measuring the actual curvature direct. Such methods impose infinitely more demands upon the stylus instrument and the data processing than the one proposed and are not thought to be viable for in-process measurement either of control of manufacture or prediction of performance.

In an attempt to devise other schemes of measurement use is increasingly being made of lasers. Obviously, the use of light for the sizing of components has been known for many years, but it is only recently that lasers have been used to assess the statistical properties of surfaces. The implication is that should this be possible then the correlation function, power spectral density (and hence peak density) would be accessible without contact and at the speed of light [6].

The most promising development has made use of the diffracting property of surfaces when illuminated at a glancing angle by a laser. The light scattered from the surface under certain conditions comprises of a number of components. One of these is the diffraction pattern produced by the finite apperture of the projection system and the other is an intensity term obtained when the light is collected by a lens by the expression

$$I\,(\theta,\alpha) = K \int_{-L_1}^{L_1} \int_{-L_2}^{L_2} z(x,y)\exp(-(j\frac{2\pi\theta}{\lambda} + k\frac{2\pi\alpha}{\lambda}))\;dxdy \qquad (11.5)$$

Where $z\,(x,y)$ is the local intensity field of the scattered light as a function of the spatial dimensions x and y and K is a constant.

The value of $I(\theta,\alpha)$ can be measured in the back focal plane of the lens to give an approximate spectral density of $z(x,y)$ which in turn, subject to suitable optical constraints, can be related closely to the two-dimensional power spectrum of the surface itself.

Thus, under certain conditions, the Fraunhoffer diffraction pattern produced by the surface is its spectrum. All that needs to be done then is to position a photodiode array under microprocessor interrogation to assess the actual pattern produced during manufacture [7]. From this control can be attempted. Although optical methods such as this are gaining ground they still suffer from the basic problem of sensitivity to debris, coolant and similar extraneous influences which will invariably cloud up a non-contacting method.

11.4 CONCLUSIONS

The techniques briefly described above only touch on the potential of surface measurement for the control of manufacture and ultimately the control of performance. This potential is becoming a reality in recent years due to

* the use of digital methods;
* the application of random process analysis to surface metrology;
* the use of coherent optics.

11.5 <u>REFERENCES</u>

1. WHITEHOUSE, D.J.:'Surfaces; A Link Between Manufacture and Function', <u>Proc.I.Mech.E.</u>, 1978, v.192, No.19, p.179.
2. GOYLER, J.F.W., SHOTBOLT,C.R.:'Metrology For Engineers' (Cassell, 1969).
3. DUTSCHKE, D.:'In-Process Measurement Of Ground Surfaces', (M.T.D.R., 1976).
4. WHITEHOUSE, D.J.:'Some theoretical aspects of error separation techniques in surface metrology', <u>Journal Of Physics E Scientific Instruments</u>', 1976, v.9, p.536.
5. WHITEHOUSE, D.J., PHILLIPS, M.:'Two-Dimensional Discrete Properties Of Random Surfaces', <u>Proc. Royal Soc. London</u>, to be published 1981.
6. WHITEHOUSE, D.J., JUNGLES, J.:'Some Modern Methods of Evaluating Surfaces', <u>Int. Conf. on Prod. Eng.; Tokyo</u>, 1974 pt. 1.
7. HINGLE, H.:'Use of Diffraction To Control Manufacture', <u>Proc. I.Mech. E.</u>, to be published 1981.

Index

Actuators, 4
  non-linearity, 80
Adaptive control, 138, 148-151,
  157, 162
Alarms, 4, 11, 93, 125
Aliasing, 40
Autocorrelation, 200, 202, 203,
  204
Automation, 174-175, 200

Bad data, 167
Back-up systems, 14, 19, 22
Broadcast network, 114, 116, 117,
  118, 120
Bumpless transfer, 25, 28
  software, 29

CAM, 176
Cascade control, 5, 138, 140
CNC, 178, 200
Communications
  loops, 118, 120
  networks, 13, 114-130
  parallel, 112
  serial, 112, 121
Concurrency, 90, 132
Central processing unit, 88-90,
  92-93, 95
Control algorithms
  coefficient quantisation, 49
  compensation of time and
    frequency domain, 57
  dead-beat, 53
  incremental, 68, 70
  jacketed, 145-149, 150
  model following, 150
  multi and sub-rate, 52
  optimal, 72, 147-151
  PID, 4, 137-138, 144, 148,
    149, 150, 151,157
  regulator, 66, 71

servo-tracking, 66
synthesis, 64, 66-75

Daisy chain, 98, 101
Data types, 132
Dead time, 102, 141
  fractional, 63, 77
Debugging, 93, 94
Decomposition, 170
Digital integrator, 79
Digital filters
  noise suppression, 3
  notch, 48
Direct digital control, 155
  benefits, 155, 156-158
  implementation, 4-7, 19,
    158-165
Distributed control, 13, 16,
    112-114, 163
DNC, 178

Economics of computer control, 15
Embedded computers, 131
Error steady-state, 54
Exception handling, 132

Feedforward, 6, 73, 138, 140, 142,
  146, 147
Filtering
  Kalman filter, 65, 138, 145,
    149
  low-pass, 48
Frame, 123
Functional modelling, 170, 179-181

Gain/phase margin, 41, 46, 57

Handshaking, 24, 26
Hierarchical systems, 12, 16, 88,
  117, 178

Identification
 extended least squares, 76
 forgetting factor, 65, 77
 least squares, 64-66
 maximum likelihood, 76, 85
 time-varying systems, 77
Information model, 179-181
Information systems, 171, 172
Information transfer, 90-92, 94,
 113, 171
Input/output, 88, 90, 103, 104
 buffered, 91
 direct, 91, 94
 intelligent, 92
 operator, 91
 process, 21, 22, 91
 programmed, 90
Instructions, 89, 96, 99
Integral windup, 5, 29, 147, 168
Integrity, 14-15, 31, 67, 81,
 114-118, 162
Intelligent modules, 92, 112
Interaction, 34-37
Interrupts, 92
 context switching, 96
 masking, 102
 multi-level, 90, 97
 non-maskable, 102
 polling, 93, 97, 98, 101
 priority, 99-102
 response, 94, 95
 software, 94, 95, 167
 vectors, 97-98, 99

Lag compensation, 47
Layered networks, 121-127
 interface, 122
 protocol, 122-123
Lead compensation, 46, 57
Lead/lag compensation, 32

Maintenance, 113, 135, 156
Manual operation, 93, 112
Manufacturing systems, 170
 batch, 173-174, 175
 flowline, 174
 job, 173
Measurements
 machine components, 201
 process, 165, 197
 surface finish, 201
 wear, 197-199
Mesh network, 115
Messages, 123
Microprocessor, 14, 15, 16

Model reference adaptive control,
 81
Modularity, 88, 92, 113, 125,
 132-134, 165
Monitors, 133
Multi-processor systems, 106, 113
Multi-terminal systems, 110
Multivariable control, 77, 160,
 162-163
Mutual exclusion, 133

Network architecture, 124-127, 129
Network configurations, 114
Network standards, 128-130
Noise arithmetic round-off, 51
Non-minimum phase, 69, 72, 78

Operating systems, 102-110, 145,
 165
Operator communication, 165
Optimization, 157, 171, 172

Packets, 123
Performance criteria, 144
Plant operator, 10-11
Pole assignment, 67
Ports, 125
Power failure, 93
Precompensation, 32, 79
Predictor, 72, 73, 76
Process engineer, 9
Programming
 interrupt, 94, 102
 real-time, 131, 165-166
 re-entrant, 102
Programming languages
 ADA, 135-136
 CORAL, 131-132, 166
 CUTLASS, 155, 165-168
 MODULA, 132-135
 PASCAL, 132
 RTL/2, 131-132, 145
PROWAY, 128-129
Pseudo systems, 72, 76

Real-time clock, 94
Recursive algorithms, 64, 65,
 75, 80
Reliability, 21, 113, 156
Ripple performance, 52, 57, 58
Robustness, 71, 78, 81

Sampling frequency, 39
Self-tuning
 extended, 78

identification, 64-66
multivariable, 77
non-parametic, 79
predictions, 80
property, 75
synthesis, 66-75
time-varying parameters, 77
Semaphores, 134
Sequence control, 7-8, 15
Software engineering, 80 131,
132
Spectral density, 200, 202, 204
Spooling, 104, 108
Standby, 19, 163
Star network, 116, 117
Storage
auxiliary, 90
fast access, 90
Store and forward, 114, 116, 118,
120, 124
Structural resonances, 48
Supervisor routines, 105
Supervisory control, 8, 19, 80
Synchronous communication, 93, 127

Tasks, 103, 125
common, 106, 133
multi-tasking, 104, 133,
166-167
priority, 107, 126
rolling, 107
synchronization, 133, 134
Three-term controllers, 7
Transducers, 2, 21
Transmission errors, 93, 112, 113,
126, 128
Transport layer, 125
Tuning, 58, 147, 148, 151
three-term controllers, 6, 9,
10, 18, 19, 28, 38
loops, 33
polynomial, 74

Wordlength, 49, 64, 89

Zero-order hold, 63